River Water Sharing

River Water Sharing

Transboundary Conflict and Cooperation in India

Editors
N. Shantha Mohan
Sailen Routray
N. Sashikumar

LONDON AND NEW YORK

First published 2010
by Routledge

2 Park Square, Milton Park, Abingdon, Oxfordshire OX14 4RN
711 Third Avenue, New York, NY 10017

Routledge is an imprint of the Taylor & Francis Group, an informa business

First issued in paperback 2018

Transferred to Digital Printing 2010

Copyright © 2010 N. Shantha Mohan, Sailen Routray and N. Sashikumar

Typeset by
Star Compugraphics Private Limited
D–156, Second Floor
Sector 7, Noida 201 301

All rights reserved. No part of this book may be reprinted or reproduced or utilised in any form or by any electronic, mechanical, or other means, now known or hereafter invented, including photocopying and recording, or in any information storage or retrieval system, without permission in writing from the publishers.

Notice:
Product or corporate names may be trademarks or registered trademarks, and are used only for identification and explanation without intent to infringe.

British Library Cataloguing-in-Publication Data
A catalogue record of this book is available from the British Library

ISBN: 978-0-415-55155-7 (hbk)
ISBN: 978-1-138-38036-3 (pbk)

Contents

List of Tables, Figure and Maps   vii
Preface   ix

Section I: Introduction

1. Locating Transboundary Water Sharing in India   3
   N. Shantha Mohan
2. The Water Sector in India: An Overview   23
   Sailen Routray

Section II: Institutional Framework

3. Boundaries of Transboundary Water Sharing   47
   Narendar Pani
4. Resolving River Water Disputes in India: Reflections   66
   Ramaswamy R. Iyer
5. Regulatory Aspects in Water Resources Development and Management   81
   R. Jeyaseelan

Section III: Historical and Technological Perspective

6. Kaveri in its Historical Setting   99
   S. Settar
7. The Cauvery Tribunal Award from a Hydrology Perspective   108
   Rama Prasad

Section IV: Negotiated Approaches and Alternative Paradigms

8. Evolving a Negotiated Approach to Sharing of Transboundary Rivers   121
   Vijay Paranjpye

9. Negotiation Through Social Dialogue:
   Insights from the Cauvery Dispute 140
   S. Janakarajan

10. Integrated Water Resource Management (IWRM):
    An Alternative Paradigm 156
    Anitha Kurup

About the Editors 181
Notes on Contributors 183
Index 187

List of Tables, Figure and Maps

Tables

7.1	Water Budget above Mettur	110
7.2	Monthly Break-up of Release at Karnataka–Tamil Nadu Border (in TMC)	111
7.3	Return Flow on the Basis of Evapotranspiration	111
7.4	Percolated Water	114
10.1	Selected Examples of National Water-Related Disputes	164

Figure

7.1	Comparison of Flows Recorded at Biligundlu and Mettur	117

Maps

7.1	Cauvery River Basin, Irrigation and Hydroelectric Projects	109
10.1	Nile River Basin	159
10.2	Mekong River Basin	162
10.3	Indus River Basin	166
10.4	Cauvery River Basin	167

Preface

River-water sharing, particularly from a transboundary perspective, is embedded in conflicts and disputes that surround fair and equitable access to, and control over common water resources. These conflicts are all-pervasive and may take place between nations, regional states, or competing sectors or users. Within India, majority of the river basins are geographically spread across two or more states, hence resulting in tensions, and at times even violence, at the inter-state and local level. To address these conflicts several national legislations such as the River Boards Act, the Interstate Water Disputes Act and others relating to environment protection, forest conservation, pollution control, etc., have been enacted. Further, the central government is also empowered to take measures to ensure integrated development of inter-state rivers, and establish mechanisms for resolution of conflicts between riparian states. Despite these efforts, they have been largely ineffective in resolving disputes between the contesting states. Many of the problems are due to the fact that there are no acceptable criteria that determine what is fair and equitable in terms of water allocations. More importantly, there is an absence of clear legal and institutional mechanisms to address these problems. Often decisions regarding allocation of water are informed by the dominant discourses of science and technology, with the exclusion of perspectives from other disciplines such as the social sciences and humanities, which are important in capturing the issues that are human in nature. Thus, a multidisciplinary perspective, adopting the tools and approaches specific to the various disciplines, is necessary to identify and evolve criteria for transboundary water sharing. Besides, the existing legal and institutional mechanisms need to be reviewed and alternatives evolved, or the existing ones need to be strengthened, for conflict resolution.

A national consultation on 'Interstate Transboundary Water Sharing in India' was held on 26–27 June 2007 at the National Institute of Advanced Studies, Bangalore. This was a platform for the different stakeholders, across political and administrative boundaries, to exchange information and learn from various perspectives on the sharing of river water. The paper presenters were drawn from diverse fields of expertise spanning grassroots leaders, academics, administrators, social scientists, historians, technocrats and gender specialists. The main aim of the dialogue between the stakeholders was to evolve a set of agreeable criteria that states could use to determine equitable allocations and also strengthen, and/or evolve alternate mechanisms for conflict resolution.

This book captures the importance of multidisciplinary approaches in arriving at a more comprehensive understanding of the complexities of inter-state water sharing, and the difficult problems faced by both states and stakeholders in addressing conflicts, particularly at the basin level. The volume is divided into four sections. The introductory chapters locate transboundary water sharing within discussions on the water sector in India. The second and third sections analyse issues from an institutional, legal, historical, cultural and technological perspective. The fourth section presents solutions in addressing problems related to transboundary rivers, and discusses an alternative paradigm for integrated water resource management.

We are grateful to all those committed people; natural and social science academics, policy makers, the government, civil society, farmer groups and community practitioners who participated in the consultation and provided deep insights. The articles have been further strengthened by the comments and suggestions of experts including Dr Roopa, Prof. Chandrakanth, Prof. P.P. Mujumdar, Prof. R.S. Deshpande, Dr D. Nagesh Kumar, Dr Nagaraj and Dr Smita Mishra Panda. We thank the distinguished authors for contributing chapters to the book, as well as for their sustained interest throughout the editorial process. We sincerely thank Dr K. Kasturirangan, Member of Parliament and Director, National Institute of Advanced Studies (NIAS), Bangalore for his continuous support in shaping the dialogue and reiterating the importance of this book for a wider audience. Our thanks to the faculty, staff and doctoral students at NIAS for their contribution in bringing out this book.

A special note of thanks to Global Water Partnership, Stockholm and Dr Mercy Dikito-Wachtmeister for extending financial and technical support, as well as, the representatives of the Global Water Partnership–South Asia and the India Water Partnership for their interest and involvement at all stages of the dialogue. Finally, we would like to express our thanks to Ms Omita Goyal, Ms Rimina Mohapatra and Ms Pallavi Narayan from Routledge for their patience and tireless effort in the publication of this volume.

N. Shantha Mohan
Sailen Routray
N. Sashikumar

Section I
Introduction

1
Locating Transboundary Water Sharing in India

N. Shantha Mohan

Introduction

The sharing of rivers across political boundaries is an area of both contention and conflict, be it at the international, national, regional or local levels. Conflicts over water have steadily increased in number due to the various reasons of population growth, rapid industrialization, consumerism, pollution, environmental degradation, inequities in the access to and use of water, poor governance and complications arising out of managing multiple uses across multiple users. The intensity of these conflicts has also increased where the water has been a scarce commodity and a diminishing resource. The increasing pressures on the existing availability of water have created a 'crisis of unprecedented proportions' (Barker et al. 2000).

India is the largest country in South Asia; the second largest country in the world in terms of population; and the seventh largest in terms of its geographical area. It has an area of 3,287,590 sq. km and its population stands at 1,027,015,248, according to the 2001 census. India has one of the largest arable areas in the world with 183.95 million hectares (ha), or about 56 per cent of the total geographical area under cultivation. Agriculture contributes about 22 per cent of the Gross Domestic Product (GDP), and employs around 65 per cent of the labour force.

India has a monsoonal climate with the south-western monsoon in summer and the north-eastern monsoon in winter contributing around 70–95 per cent of the annual rainfall. The average rainfall for the whole country stands at 1,170 mm with a lot of spatial variation. The annual rainfall is uneven and varies from lower than 150 mm/year in the deserts of north-western Rajasthan to more than 10 meters/year of rainfall on the Khasi hills in the north-east. Further, most of the country receives rain for only 100 hours in a year, and half of that total in less than about 20 hours (Agarwal and Narain 1999).

Monsoonal rainfall and the melting of the Himalayan glaciers in the summer months are the two main sources of water in India. The rivers of India can be grouped under four categories; the Himalayan rivers, the rivers of the Deccan plateau, the coastal rivers on the west coast and the rivers of the inland drainage basin in western Rajasthan. Of these only the Himalayan rivers are perennial and account for most of the surface water flows of the country. They include Indus, Ganga and the Brahmaputra. Other rivers in peninsular India such as the Narmada, Godavari, Krishna and Cauvery are completely dependent on the rains. All major rivers including the Indus, Ganga, Brahmaputra, Narmada, Mahanadi, Godavari, Krishna and Cauvery, are inter-state rivers insofar as they originate in one state and flow through another state or states before reaching the sea. Several smaller rivers in the country are also inter-state rivers (Wood 2007).

According to the National Commission on Integrated Water Resources Development Plan (NCIWRDP, 1999) the total available surface water resources in India are estimated at 1,953 billion cubic meters (BCM); out of this only 690 BCM is utilizable. The available and utilizable groundwater resources are put at 432 BCM and 396 BCM, respectively. The country's total water requirement by 2050 of both surface and ground water is estimated at a high and low of 1,180 BCM and 970 BCM, respectively.

Cooperation over Transboundary Rivers

International Experience

The sharing of river waters across political boundaries is a matter of conflict in many countries. All of the 264 largest rivers in the world flow through basins that are shared by more than one nation and home to at least 40 per cent of the world's population (Wolf 1998). 'Water' and 'war' together are two topics being talked about with increasing frequency. The basic argument for 'water wars' is that water is an essential resource, and the scarcity of water in an arid and semi-arid environment leads to intense political pressures, often referred to as 'water stress.' Although 'water wars' might be mythical, the connection between water and political conflicts is very much real. Most water-related conflict and violence happens at the sub-national level; between various social groups, various water-use sectors, or between the various administrative units. While no 'water wars' have occurred, there is enough evidence that shows that the lack of adequate freshwater has led to occasionally intense political instability that might result in violence, sometimes acute, on a small scale; what this means is that the geographic and political scale and the intensity of conflict are inversely related to each other. It is observed that the absence of proper dispute resolution mechanisms can lead to 'water wars.' A close examination of the case studies cited in such literature shows that there has never been a single war fought over water alone (Wolf 1998).

The history of its resolution, in contrast to that of water conflict, is much more impressive. Over 3,600 treaties have been signed historically over different aspects of international waters, out of which almost 150 have been signed in the last century that deal specifically with access to water control. These cooperative regimes are also very resilient over time, even when the riparians are otherwise hostile to each other.

Examples of transboundary water-sharing from around the world reveal that various institutional mechanisms in the form of committees, councils, commissions and regulatory boards have been established to facilitate sharing of waters and conflict resolution through cooperative and coordinated approaches. Some of the notable examples in international cooperation in river basin water sharing are the Lower Mekong River Basin and the Nile. Both the basins have created institutional spaces that promote the member nations to coordinate and contribute to regional water resource development in spite of environmental, political and economic constraints and tensions surrounding the basin.

The reasons for their success in promoting cooperation have been in their efforts to build confidence amongst the member countries at practical levels by focusing on common interests such as small-scale programmes involving interventions and studies, generation of integrated basin level database including geographical data from maps and satellite pictures, hydro-meteorological data and socio-economic data. This is especially important since most available information on the basins are either unavailable or inadequate.

An example of cooperation in the context of inter-state transboundary water sharing in a federal system is the Murray-Darling River Basin in Australia. Responding to the crisis of serious conflicts amongst the concerned states, innovative reforms were made institutionalizing the Ministerial Council and the Basin Commission. While the Commission addressed matters related to technology and administration, the Ministerial Council addressed those related to policy. Though the decisions by these institutions are not legally-binding due to the consensual approach adopted by them they still enjoy a certain degree of political legitimacy.

Experience in India

India has a large number of rivers that traverse international and inter-state boundaries as well as face conflicts on common water resources. The Indus Water Treaty between India and Pakistan is an example of a successful resolution of a transboundary river water dispute. The resolution was possible on the premise that the entire Indus basin is a single unit with sufficient water to meet the demands of both countries, and that the negotiations were based on a sound technical rather than a political perspective. The final award of the treaty allocated nearly 80 per cent of the water of the entire Indus system to Pakistan and 20 per cent to India. Despite criticism, the treaty has survived three wars and five decades between the countries and remains an example of successful transboundary water sharing in a politically volatile region.

The long standing dispute between India and Bangladesh over the waters of the Ganges and the Farakka barrage was addressed when a 30-year treaty was signed between the two countries. The treaty is a step forward towards a satisfactory solution to water sharing acceptable to both countries. However, it did not include matters relating to management, conservation and prevention of pollution of their common rivers. This treaty apart from being a water-sharing treaty also lays down a framework regarding the principles for future agreements on other transboundary rivers between the two countries.

India also has a significant number of inter-state transboundary water sharing based on the principle of cooperation. They include the rivers Sutlej, Ravi-Beas, Tungabhadra, Chambal, Damodar, Gandak, Parambikulam-Aliyar, Subarnareka and Jamni. Even in a complex multibasin and multipurpose (irrigation, hydro power, domestic and industrial water supply) project like Parambikulam-Aliyar, the establishment of a Joint Water Regulation Board comprising of members from the riparian states, the cooperative approach adopted to facilitate mutual benefit has been successful. However, currently many of them are facing serious situations of conflict.

The above examples describe the importance of an integrated and cooperative approach to river basin planning, development and management. They emphasise the need to establish institutional mechanisms that promote the participation of member nations and states in generating basin-level database, conducting joint studies, evolving policies and interventions for mutual benefit.

Water Use and Water Institutions in the Indian Context

Water as a resource has multilayered competing demands and contending uses. Agriculture is by far the biggest user of water followed by industrial and domestic use. Irrigation supports nearly 60 per cent of agricultural production although irrigated lands are only a third of the total arable land; the extension of irrigation has also helped to extend crop diversification. Despite growing productivity and intensity of cultivation the irrigated crop yields are still relatively low compared to other developing countries; this is a result of poor water management.

The ultimate irrigation potential of the country according to the definitions adopted by the Government of India stands at 113.5 million ha; out of this 58.5 million ha is supposed to be created from major and medium irrigation schemes and 55 million ha from minor schemes. In 1993, the area equipped for irrigation was around 50 million ha. Irrigation is mainly concentrated in the northern states of the country with the irrigated lands being fed from the perennial rivers of the Indus and Ganges river systems. The average overall water use efficiency in canal irrigation systems is estimated at 35–40 per cent (NCIWRDP 1999). A major problem plaguing large surface irrigation systems in India is the lack of proper drainage facilities that results in salination and water logging. Various estimates point at the fact that around 6 million ha of irrigated land, around 10 per cent, of the net irrigated land in India is affected by either salinity or water logging.

There has been an explosive growth of ground water irrigation during the latter half of the 20th century. Irrigation from ground water sources is found to be more than 50 per cent in most Indian states. While there has been considerable development in addressing issues regarding sharing of surface waters it is lacking in ground water sources. For instance, United States and Mexico have several treaties governing the waters of the border regions, yet these treaties are silent on ground water with potentially disastrous results (Aparicio and Hidalgo 2004). Therefore, it is increasingly becoming an area of stress and conflict. Thus, ground water remains an area of contention despite there being consensus that ground water and surface water be subject to the same rules as they both belong to a single hydrological cycle.

Water to meet basic human needs, particularly 'water for life' is of paramount importance to enhance the quality of life. The basic water requirement has been estimated at 50 litres per capita per day (lpcd) which includes water for drinking, sanitation, cooking and bathing (Gleick 1996). In rural and urban water supply, the tendency is to project future needs on the basis of per capita norms which are fairly high: 140 lpcd in urban areas and 70 lpcd in rural areas (Iyer 2007). A projected estimate of 200 lpcd and 150 lpcd in urban and rural areas, respectively has been put forth by the NCIWRDP, while a common norm of 100 lpcd has been identified as adequate for both urban and rural areas (Iyer 2007).

The quality of water is as important as the quantity of water available and accessible to the competing users. The primary uses considered are parameters which relate to adequate and safe drinking water, water for livelihood, water for the health of ecosystems and water free from pollution.

Under the Constitution of India (apart from the important subject of Inter-State Rivers) water is the responsibility of the States and they are responsible for the planning, implementation and management of water-related projects. The apex body for the planning and management of the water resources in the country is the Ministry of Water Resources (MOWR). Central organizations concerned with the water sector such as the Central Water Commission (CWC), the Central Groundwater Board and the National Water Development Agency are under the MOWR. Their mandate is to provide technical support whereas inputs in terms of research and training are the mandate of organizations such as Water and Land Management Institutes, various agricultural universities and institutions. The Planning Commission has a critical role to play at the national level as it oversees project clearance and financial allocation to water projects. Other central agencies having a stake in the water sector include the ministries of economic affairs, agriculture and rural development, environment and forests, housing and urban development, and health.

There have been conscious attempts to create organizations to foster inter-state and centre–state coordination in the water sector. These include river boards having the responsibility of coordinating water allocation of important inter-state rivers, the National Water Resources Council (NWRC) and the National Water Board (NWB). The NWRC (set up in 1983) is the apex body in the water sector in India. It is chaired by the prime minister and has the union minister of Water Resources and the chief ministers/lieutenant governors of all states and union territories as its members. The NWB (set up in 1990) is generally seen as the executive branch of NWRC. It is chaired by the secretary of the MOWR, and has the chief secretaries of the states/union territories, and the secretaries of the relevant union ministries and the chairman of CWC as its members. Local government institutions such as municipalities and panchayats are important institutions in the water sector due to their mandated role in ensuring drinking water supply. Pollution control boards that generally operate under the ministry of environment and forests (at both the central and the state levels) are responsible for water quality.

Rules and Mechanisms to Address Transboundary Water Sharing

The International Experience

Before making an attempt to evolve mechanisms to resolve disputes surrounding the waters of inter-state rivers it might be instructive to explore the reasons for the genesis of such disputes. Most conflicts over river waters develop around large water resource development projects as more often than not they involve substantial changes to the status quo in terms of water usage patterns that have evolved over generations, and they tend to ignore the issues surrounding political power and equity. Therefore, there is an urgent need to rethink the paradigm of water resource development in the country.

River water sharing and conflict resolution between states through a workable compromise formulae are constrained by inadequate information and database, ineffective institutional mechanisms, hardened regional identities and loyalties, and the threat of economic hardship. The political boundaries that divide states, which is a political construct, often subsume issues that are humane, common and social in nature. Many of the problems are due to the fact that there is an absence of a set of acceptable parameters and indicators that form the basis of conflict resolution, surrounding issues of transboundary water sharing. While it is not possible to proclaim absolute parameters uniformly for all states and river basins, it is possible to identify those parameters that are non-negotiable and can be applied in all cases along with the parameters specific to river basins. The uniqueness of each basin and its riparian states suggest that any universal set of principles must, by necessity, be fairly general (Wolf 1999).

The Helsinki Rules and the UN Convention of 1997 provide the principles of equitable and reasonable sharing for beneficial uses as the basis for sharing transboundary river waters. Further, there is the acceptance of the principle of no 'substantial' or 'significant' harm should be caused to the lower riparian by the upper riparian. The UN Convention, however, limits its rules to transboundary contexts. The Helsinki Rules and the UN

Convention define reasonable and equitable use based on a non-exhaustive list of seven relevant factors. In Article 6 of the ILC it is suggested that 'the weight to be given to each factor is to be determined by its importance,' and that 'all relevant factors are to be considered together.' However in Article 10 it is stated that, 'in the event of a conflict between uses ... [it shall be resolved] with special regard being given to the requirements of vital human needs.' While the Helsinki Rules deal with issues concerning pollution, navigation, timber floating, the UN Convention fails to integrate the environmental and relevant human rights concerns.

The claims and counterclaims among states involved in disputes over surface waters follow a set pattern that diverge sharply according to the riparian status of the state (Dellapenna 2006; McCaffrey 2001). While the upper riparian States claim 'absolute territorial sovereignty' the downstream states claim 'absolute integrity of the watercourse'. On the one hand, the upper riparian states claim complete control over the water within their borders without regard of its effect on the downstream states. On the other hand, downstream states claim that the quantity or quality of water that flows down the watercourse should not be affected by the upstream states. Solutions have been sought to these conflicting claims by applying the principle of 'equitable utilization' as formulated by the Helsinki Rules to express the rule of restricted sovereignty as applied to fresh waters. 'Each basin state is entitled, within its territory, to a reasonable and equitable share in the beneficial uses of the waters of an international drainage basin' (ILA 1966: Article IV). The principle of equitable utilization recognizes the right of all riparian states to use water from a common source so long as they do not interfere unreasonably with uses in another riparian state. Thus, the numerous water treaties demonstrate that each state's sovereignty over its water resources is restricted by the obligation not to inflict unreasonable injury on another state (United Nations 1997). The Helsinki Rules treat international drainage basins (watersheds extending over two or more States) as indivisible hydrologic units to be managed as a single unit to assure the 'maximum utilization and development of any portion of its waters' (ILA 1966, Article II).

Expanding the scope of the Helsinki Rules and The UN Convention, the Berlin Rules on Water Resources enumerate five general principles that apply to states in the management of all waters. They include participatory water management, conjunctive management, integrated management, sustainability and minimization of environmental harm. Additionally, it posits three further rules relating to water in a transboundary context. They are cooperation, equitable utilization and avoidance of transboundary harm (Dellapenna 2006). The most cooperative of neighbouring states have found it difficult to achieve mutually acceptable arrangements to govern transboundary surface waters even in relatively humid regions where fresh water is usually sufficient to satisfy most or all needs (McCaffrey 2001). Within federal unions and at the sub-national level the states have engaged in political and legal struggles over the surface waters they share and contest its allocation. To address these struggles, there is a need for cooperative measures with the participation of all concerned parties. In fact, evidence suggests that cooperative approaches to water scarcity problems are more likely to result in positive solutions than conflicts.

In the context of the framework provided by the guiding principles, the most important and non-negotiable parameters include water for life in terms of adequate and safe drinking water, water for livelihoods and water for environment protection and quality of water.

The Indian Experience

Water conflicts can be of many kinds depending upon the contesting parties and the nature of the contestation. One of the many types of conflicts can be inter-sectoral. The resolution of these kinds of conflicts can be located in the general framework created by water-use prioritization as specified in the National Water Policy and as implied in the Indian constitution. For the resolution of conflicts, negotiated agreements for water resource development and sharing were often the institutional arrangements resorted to. But often there were difficulties in reaching settlements through negotiations between the concerned parties and other mechanisms were adopted.

Conflicts surrounding river-water sharing raise questions related to Indian federalism. The problems relating to conflict resolution is compounded by the lack of clear legal and institutional mechanisms available at the national level. Irrigation still accounts for more than four-fifths of all water consumption and is listed under the State List in the Indian constitution. The most important entry regarding water is Entry 17 in the State List subject to the provisions of Entry 56 in the Union List that gives extensive powers to the central government to legislate regarding inter-state rivers. However, not much use has been made of the enabling provisions provided. With the growing complexity of problems on centre-state relations, a few consultative bodies were set up to review the policy and other related matters and to arrive at a consensus on different contentious issues. Therefore, water is as much a central subject as it is a state subject. Article 262 of the Constitution of India provides for a strong role for the centre for the adjudication of disputes surrounding inter-state rivers. This article has been operationalized by the Inter-State Water Disputes Act (ISWD), 1956 that provides for the setting up of tribunals for settling disputes surrounding inter-state rivers.

According to the ISWD Act, a state government which has a water dispute with another state government may request the central government to refer the dispute to a tribunal for adjudication. The tribunal consists of a chairman and two other members nominated by the Chief Justice of India, from among persons who, at the time of nomination are, judges of the Supreme Court. The tribunal can appoint assessors to advice in the proceedings, investigate the matter and make its report along with its decision. The decision has to be published and is final and binding on the parties. The awards of the tribunals set up for the Krishna, Godavari and Narmada rivers are considered as relatively successful examples of dispute resolution under this Act. But recently there seem to be problems surrounding the success of the tribunals set up under the Act to settle disputes; the Ravi-Beas case and the Cauvery dispute are illustrative as both the awards have run into political rough weather and the process of adjudication itself has been marked by intense politicking.

Despite the recent amendments to the ISWD Act giving the Tribunal's award the status of a Supreme Court decree, disputes often remain unresolved. Though the RBA provides for arbitration as an instrument of dispute resolution, governments have resorted to adjudication resulting in the overuse of the ISWD Act which has led to a lot of wastage of time and resources of the nation as a whole (Gosain and Singh 2004). Generally a lot of time is taken to reach a final settlement, even though the Act (amended in 2002) provides for a six-year time limit, a long period of time to resolve issues of such magnitude and import. Since there is an increasing tendency to avoid compliance of awards of tribunals and court orders, there is need for greater involvement of the Union government, keeping in view the national interest and protection of human rights. Inter-state water disputes have been a major irritant in the maintenance of cordial relations between riparian and non-riparian states.

Arbitration and negotiations are a better way to resolve disputes rather than adjudication which should be resorted to only in the case of failure of negotiations between the disputing parties. One such institutional arrangement at the basin and sub-basin levels is the River Basin Organization (RBO) under the centrally legislated River Boards Act 1956 (RBA) which was legislated under the provisions of Article 56 of the Union List. The RBOs are mandated to regulate and develop inter-state rivers and river valleys at the request of the riparian state. The boards can have members with expertise in various fields such as irrigation, engineering, flood control, water and soil conservation, administration and finance, thus allowing for a holistic approach to the arbitration process. However, no such river board has been established, largely because no state government was in favour of such a course (Iyer 2007).

Governance relating to water is further decentralized through the 11 and 12 Schedules of the Constitution which lay down lists of subjects to be devolved to the panchayats and nagar palikas. The List includes drinking water, water management, watershed development, sanitation and so on. These local self-government bodies that have an important role to play in relation to water distribution in an equitable and efficient manner need to be rejuvenated. However, there are very few institutions at the middle level to resolve such conflicts. For example, if disputes arise between communities within a river basin or canal system, there are very few, if any, mechanisms for resolving them in a satisfactory and participatory manner. Water bureaucracies at this level remain inaccessible. Institutional innovations at the basin or sub-basin level can play an important and useful role in promoting conflict resolution by facilitating the representation and participation of all water users in the community.

Assessing the inadequacy of the existing mechanisms to address conflicts between states, the Administrative Reforms Commission of 1969, the Rajamannar Committee of 1971 and the Sarkaria Commission of 1983 recommended the setting up of an Inter-State Council (ISC) as envisaged under Article 263 of the Constitution with the responsibility to inquire into and advise upon disputes between the states. The ISC was set up in 1990, as a federating consultative body under the chairmanship of the prime minister. The council is a recommendatory body investigating and deliberating such subjects, in which some or all of the states, or the union and one or more of the states have a common interest and make recommendations regarding relevant subjects for the better coordination of policy and action (Iyer 2002).

Given the complexities of water use within society, developing, allocating and managing it equitably and efficiently and ensuring environmental sustainability requires that the disparate voices are heard and respected in decisions over common waters and use of scarce financial and human resources. At the micro-level, there are several non-official efforts that have also been put forward by civil society to address issues of transboundary water sharing which need to be strengthened. One such example is that of the initiative undertaken by the Madras Institute of Development Studies to create a platform for dialogue between the farmers of the two contending riparian states of Karnataka and Tamil Nadu in the Cauvery Basin. Through a multistakeholder platform it initiated a dialogue between farmers which has resulted in the establishment of the 'Cauvery Family'. The members of this family have developed empathy and understanding for each others' problems and needs, and are therefore able to influence their respective governments from the perspective of the collective stakeholder.

Another example is the initiative of Gomukh in Maharashtra, a non-governmental organization involved in interventions related to land and water issues in the Bhima Basin. The Kolwan Valley experience highlights the need to bring together all stakeholders; communities, state agencies, civil society organizations to find solutions to critical water sharing problems. Negotiation has been an important element in the decision-making process. To facilitate informed negotiating positions, networks and coalitions among the stakeholders have been established and strengthened.

In both the above cases accommodation of different stakeholder interests is central to the decision-making process. It allows for an inclusive framework (institutional and administrative) within which people with different interests can discuss the polarized discourse on water (Rogers and Hall 2003). Further, it encourages a consensual framework that accommodates and appreciates different stakeholder interests and points of view including the concerns of equity and sustainability. If such a consensual framework is not evolved, 'rivers will continue to divide us, emotionally and politically, leading to a million revolts' (Gujja et al. 2006). The approach also emphasizes the importance of generating reliable data arising out of a common interest and shared vision of developing the basin, honouring each other's requests for information, dialogue and transparent procedure to participate in the negotiation processes to resolve conflicts. 'Stakeholder representatives must get formally involved in the sharing of agreements at the basin scale. This will allow the negotiations to "deepen" and the agreements to be more effectively implemented and enforced' (Vaz and Zaag 2003: 50).

An analysis of the mechanisms adopted by the state and civil society is to highlight both the strengths and limitations in addressing issues related to inter-state water sharing and conflict resolution. Disputes surrounding inter-state water sharing may occur when the basins of the rivers involved cut across political boundaries, cultural identities and livelihoods. Under the circumstances, it appears that the Inter State Council (ISC) is an important unexplored mechanism that has the potential to develop into an effective institutional mechanism to address this issue. The ISC has been envisaged in Article 263 of the Indian constitution with the mandate of enquiring into and advising upon disputes arising between the various states of India, to investigate subjects of common interest amongst the states, and to make recommendations upon such subjects for the better co-ordination of policy and action.

Despite the provision in the Constitution, and these recommendations, the ISC was set up as a recommendatory body by a Presidential Order only in May 1990. This high-powered council comprises as its members the prime minister of India, chief ministers of all states, chief ministers of union territories, administrators of union territories, six ministers of cabinet rank in the union council of ministers and permanent invitees.

Further, the Standing Committee of the ISC has been constituted in 1996 for continuous consultation and processing of matters pertaining to centre-state relations for consideration of the Council, and monitoring of implementation of the decisions taken by the Council. The Committee comprises of the home minister as its chairperson, with five union ministers of cabinet rank, and nine chief ministers as members nominated by the chairman of the ISC.

Activation of the ISC is one of the main agendas of the UPA Government at the centre in 2004, announced in its National Common Minimum Programme and reflected in the strategic action plan (SAP) developed for the same purpose. Conducting studies, establishing various subject-special forums for sustained consultation, providing an active forum for free and frank deliberation on various complex public policy and governance issues, and building consensus are some of the activities that the council has been engaged in, thus facilitating the setting up of a healthy institutional framework for resolving centre-state relations. As must be evident from this short discussion on the mandate, and the membership structure of the ISC, it has the potential to provide an alternative forum for addressing concerns surrounding equitable distribution of water across the boundaries of the various states in India without changing the present constitutional structure regarding water.

However, the existing structure and governance has to be expanded to make these fora inclusive, participative and consultative so that the interests of all stakeholders, the government, academics, policy makers, media, civil society organizations, legal experts and water users, etc., is integrated. For this, the Integrated Water Resources Management (IWRM) has emerged as a logical conceptual alternate to the sectoral, top-down management approach to deal with problems surrounding water. It facilitates inter- and multidisciplinary, inter-sectoral and multistakeholder approach to integrate the different perspectives in promoting equitable water distribution and in conflict resolution among the various users.

Organization of the Volume

This volume is divided into four sections. Section I or the Introduction comprises of two chapters. The second chapter in this section authored by Sailen Routray is titled 'The Water Sector in India: An Overview'. Here, the author provides a broad perspective of the water scenario in India by providing succinct summaries of the various developments in the sector and by discussing newly emerging alternative paradigms. While discussing the dominant discourses and the emerging approaches to water management, he points out the inadequacies of the existing legal frameworks and institutional mechanisms to manage water equitably and efficiently. He argues for a paradigm shift and policy change that prioritizes the concerns of equity, social justice and rights, particularly of the marginalized sections of society, and puts forth the need for the adoption of eclectic, innovative, and context-specific policy initiatives in the sector. This chapter, thus, provides the broad context for locating the other chapters that follow in the succeeding three sections.

Section II (Institutional Framework) consists of three chapters which focus on the institutions and organizations that enable transboundary water sharing and conflict resolution. The first chapter of the section addresses the issues surrounding the 'boundaries' of transboundary water sharing where Narendar Pani argues that while water scarcity is an area of contention, river water conflicts cannot be the result of scarcity alone. Giving illustrations, he makes a case for the need to focus on the changes that provide opportunities for cooperation, and the institutional capacity to manage such changes. Ramasamy R. Iyer in the second chapter of this section, titled 'Resolving River Water Disputes in India: Reflections', presents the existing legal and policy frameworks along with the procedural and operational dimensions of river water disputes. He illustrates the need to incorporate groundwater availability in the allocation of water to disputing riparian states. He also critically analyses the limitations of the available adjudication mechanism, and recommends reforms to make the processes involved more consultative, participative and interactive. In the third chapter of the section, titled 'Regulatory Aspects in Water Resource Development and Management,' Jeyaseelan discusses how a multitude of issues are interrelated to each other within the transboundary water sharing framework of integrated water resources development and management. He critically analyses the constitutional and other existing regulatory mechanisms available, and recommends the development of suitable mechanisms for ensuring water for all in acceptable quality and quantity.

Section III, titled as 'Historical and Technological Perspectives,' argues for taking into account both the history of water usage in a region/basin and the technological constraints imposed by the physical nature of water as a resource to be able to initiate a meaningful dialogue about transboundary water sharing. Both these chapters focus on the River Cauvery in South India, as this river basin is one of the oldest and most contentious of the major river water disputes in India. S. Settar in his chapter, titled 'Kaveri in its Historical Setting', traces the river and the developments in the river basin from a historical perspective right from the first references to it in the oldest extant body of Tamil literature, *Sangam*, and the first Kannada literary work called *Kavirajamarga*, to the constitution of the tribunal to resolve conflicts between the disputing states of Karnataka and Tamil Nadu.

In the chapter titled 'The Cauvery Tribunal Award from a Hydrology Perspective', Rama Prasad throws light on some of the contentious hydrological observations like yield determination, releases, return flow, percolation loss in paddy cultivation, groundwater and so on. He also discusses the implications and the problems in the implementation of the final award of the Cauvery Tribunal, and their non-amenability to solutions at the Regulatory Authority level.

Section IV of the book is titled 'Negotiated Approaches and Alternative Paradigms', and contains three chapters. In the chapter on 'Evolving a Negotiated Approach to Sharing of Transboundary Rivers' Vijay Paranjpye makes a case for creating a 'participatory-democratic-culture,' and discusses the role of negotiation in reaching agreements. Illustrating with an example of the Bhima River Basin, he emphasizes the importance of administrative transparency, accountability, mutual trust, respect and responsibility among the stakeholders, the authorities, and the negotiators for resolving disputes. The chapter titled 'Negotiation through Social Dialogue: Insights from the Cauvery Dispute' by Janakarajan presents the process involved in creating a platform for dialogue primarily between farmers, and traces the experience of the 'Cauvery Family'. He also discusses the role of the Cauvery Family in developing a feeling of trust among the farmers, redefining the issues of conflict in the larger socio-economic and cultural context, and in facilitating the process of resolving the dispute by adopting a scientific approach with a humane perspective. The need for an integrated and holistic approach is discussed by Anitha Kurup in the chapter on 'Integrated Water Resource Management: An Alternative Paradigm'. In her critique of the existing practices of IWRM, she discusses its limitations and argues for IWRM practices that incorporates the principles of equity and social justice. Viewing water conflicts as a part of inequities in power relations, she makes a strong case for the inclusion of women and the poor in the negotiation process. She also emphasizes the importance of drawing on the indigenous knowledge systems to address scarcity. All the four sections together argue for creating multi-dimensional and multi-disciplinary frameworks for locating interventions in the water sector so as to facilitate more equitable and efficient sharing of water between the various states in India.

References

Agarwal, Anil, Sunita Narain and Srabani Sen (eds). 1999. *The State of India's Environment: The Citizen's Fifth Report*, No. 5. New Delhi: Centre for Science and Environment.

Aparacio, Javier and Jorge Hidalgo. 2004. 'Water Resources Management at the Mexican Borders', *Water International*, 29, pp. 362–74.

Barker, Rudolf, Barbara Van Koppen and Tushar Shah. 2000. *A Global Perspective on Water Scarcity/Poverty: Achievements and Challenges for Water Resources Management*, Colombo: International Water Management Institute.

Cauvery Water Disputes Tribunal. 2007. 'The Report of the Cauvery Water Disputes Tribunal with the Decision', vol III. New Delhi.

'Criteria for Equitable Allocations: The Heart of International Water Conflict,' 1999. *Natural Resources Forum*, 23(1): 3–30.

Dellapenna, Joseph W. 2006. *The Berlin Rules on Water Resources: The New Paradigm for International Water Law*. Villanova: University School of Law.

Gosain, A.K. and K. Singh. 2004. 'Water Rights in Indian Transboundary Watercourses', *Jalvigyan Sameeksha*, 19(1–2): 51–60.

Gujja, Biksham, K. J. Joy, Suhas Paranjape, Vinod Goud, Shruti Vispute. 2006. 'Million revolts' in the Making', *Economic and Political Weekly*, 41(7): 57–74.

International Law Association (ILA). 1966. 'The Helsinki Rules on the USCO of Waters of International Rivers', in *Report of the 52nd Conference–Helsinki*.

Iyer, Ramaswamy R. 2002. 'Inter-State Water Disputes Act 1956: Difficulties and Solutions', *Economic and Political Weekly*, 13 July: 290–910.

———. *Towards Water Wisdom: Limits, Justice, Harmony*. 2007. New Delhi: Sage Publications.

Jain, S.N., Alice Jacob and Subhash C. Jain (eds). 1971. *Interstate Water Disputes in India: Suggestions for Reform of Law*. Bombay: N.M. Tripathi Pvt Ltd.

McCaffrey, Stephen. 2001. *The Law of International Water Resource: Non-Navigational Uses*. Oxford: Oxford University Press.

Mohan, N. Shantha, N. Sashikumar, Sailen Routray and K.G. Sreeja. 2007. *National Dialogue to Review and Evolve Parameters for Interstate Transboundary Water Sharing in India*. NIAS.

National Commission on Integrated Water Resources Development Plan (NCIWRDP Report). 1999. New Delhi: MOWR.

Rogers, Peter and Alan Hall. 2003. 'Effective Water Governance', GWP Technical Committee.

Rowland, Marty. 2005. 'A Framework for Resolving the Transboundary Water Allocation Conflict Conundrum', *Ground Water*, 43(5).

United Nations. 1997. *UN Convention on the Non-Navigable Uses of International Water Courses.*

Vaz, Alvaro Carmo and Pieter van der Zaag. 2003. *Sharing the Incomati Waters: Cooperation and Competition in the Balance.* UNESCO.

Wolf, Aaron T. 1988. 'Conflict and Cooperation along International Waterways', *Water Policy*, 1(2).

Wood, John R. 2007. *The Politics of Water Resource Development in India: The Narmada Dams Controversy.* New Delhi: Sage Publications.

2
The Water Sector in India: An Overview
Sailen Routray

Introduction

India, with around 16 per cent of the world's population and 2.45 per cent of the world's land area, has only 4 per cent of its water resources. In gross national terms the availability of water is not uncomfortable right now. This situation can change with population growth, economic development and increasing urbanization. According to the National Commission for Integrated Water Resources Development Plan projections, supply will fall short of the high demand in the future. The commission believes that this will make for a difficult situation but a crisis is certainly avoidable if timely measures are taken to balance demand and supply. National level aggregates of course pose many problems. India, as the cliché goes, is a land of wide variations and this extends to the water sector as well. In terms of water availability there exists a wide range of both spatial and temporal variations. Most of the precipitation in the country happens through rainfall, most of which is received in the four months between July and October in most part of the country as a result of the south-west monsoons. Spatially, broadly speaking, the northern and eastern parts of the country have better water endowments compared to the western and southern parts that are relatively less endowed. The latter include the desert areas of Rajasthan and arid areas in the states of Gujarat, Maharashtra, Karnataka, Andhra Pradesh and Tamil Nadu in peninsular India that lie in one rain-shadow region or the other (Iyer 2003).[1]

Sectoral Usage of Water

Although water availability in India is spatially and temporally variable, there is a lot of consistency in terms of the pattern of demand. Irrigation is the biggest consumer in all Indian regions accounting for more than 80 per cent of all water consumed. Domestic usage (including drinking water) and industrial usage together account for less than 20 per cent of all water demand (Iyer 2001a). Since irrigation is the single biggest consumer of water in the country, the sector needs to be looked into in some detail.

Irrigation

Generally, the discourse surrounding irrigation in India gets shaped around surface irrigation, though ground water is an increasingly significant contributor. Over the last three decades or so, India has seen a massive expansion of ground water irrigation. There has been a progressive decline in the water table in most areas, and this has led to the farmers deepening wells in a competitive manner. This has also resulted in increasing costs of irrigation that is being increasingly borne by the poorer farmers. Urban water demands for both domestic and industrial purposes have spiralled upwards during the same period of time, and a bulk of this demand has been met through the exploitation of ground water in urban areas. Lack of treatment of effluence has contributed to the pollution of the existing ground water stock. The degradation of the ground water through over-extraction and pollution is increasingly contributing to inequity, conflicts, indebtedness and poverty (Janakarajan and Moench 2006).

Surface irrigation projects were very important in providing irrigation water until about 1970; starting with the early seventies the rate of expansion of ground water irrigation has been higher than that of surface water irrigation. The expansion of ground water has sometimes been able to extend the benefits of irrigation to areas hitherto unreached by surface irrigation systems. It is also a much more efficient source of irrigation as there are fewer evapo-transpiration losses, and it allows farmers better control of both the timing and quantity of water. But due to the unsustainable growth over the last decade or so, the ground water irrigation economy might end up collapsing under its own weight (Shah 2004).

In terms of the existing surface-based systems there have been no major initiatives to improve the efficiency of water use through renovations. Till now the attempts at promoting water-use efficiency by inducing institutional changes have not been very concerted, and have not had a major impact in terms of achieving goals. In India, the proportion of the irrigated area transferred to the Water Users Associations (WUAs) is around 7 per cent; this compares very unfavourably to 45 per cent achieved in Indonesia, 66 per cent in Philippines and 22 per cent in Thailand. Although, in the long-term, institutional changes may hold greater promise, in the short-term, technological changes resulting in improved crop yields per unit of water, inducing farmers to switch to water-efficient crops by pricing mechanisms; the use of rain water in conjunction with irrigation water in high rainfall areas, and greater focus on the high rainfall areas of central and eastern India might be more promising as strategies. Increasingly the role of the government in irrigation systems is seen as a part of the problem; this leads to the prescription of market mechanisms as the cure. But as experience has more often than not shown, market failures in water and irrigation management are as common as the failure of state systems. There is a widespread prevalence of externalities, high transaction costs, lack of clear property rights for irrigation water in India. All this make it difficult to have properly functioning water markets and any large-scale, full-scale privatization of irrigation water difficult and potentially inequitable (Hanumantha Rao 2002).

Drinking Water

Although irrigation is the biggest consumer of water in India, the most important usage of this resource is for drinking and other domestic purposes. Despite the gains made in the water and sanitation sector after Independence, tens of millions of people in India are yet to gain access to safe drinking water. About 15 lakh children under the age of five die every year in the country due to preventable, water-borne diseases, and about 20 crore persondays are lost annually (mostly by the poor and the vulnerable) as a result of such diseases. In 1991, urban water supply coverage was 83.63 per cent which was an increase of only 10 per cent points over the figure in 1981. Instead of trying to achieve international standards in terms of water and following the targets of total sanitation, various governments have chosen to lower standards, with the five-year plan documents periodically revising targets. The Eighth Five-Year Plan (1992–97) set the target of 100 per cent coverage of safe drinking water in urban areas by 2000. After this plan period, the approach paper to the Ninth Plan (1997–2002) claimed that only 85 per cent of the urban population had access to safe drinking water, and admitted to inadequate coverage in slums and other poorer localities. The distribution of water in urban areas is extremely inequitable and is biased against the poor. The leakage of water in the big cities of India averages around 35–40 per cent; reducing leakages can augment supplies with minimal additional investment (Ramachandraiah 2001).

One of the key players in the emerging policy consensus surrounding urban drinking water all across the world has been the World Bank that has been promoting public–private partnership models that are supposed to be based on management contracts. In most of such contracts, a majority of the risks are supposed to be borne by the governments (primarily in developing countries) while the companies more often than not do not invest anything. A similar model was adopted by the Delhi state government recently for pursuing so-called reforms in the water sector. This was met with widespread protests, and led to the unconditional withdrawal of the loan application by the Delhi Water Board to the World Bank (Bhaduri and Kejriwal 2005).

Despite the commitments of successive governments for providing full drinking water coverage to all rural habitations, many villages still lack access to safe drinking water. A lot of the so-called no-source villages that are provided with drinking water again revert back to no-source status. This may be because of biophysical factors. But it has to be kept in mind that 'no-source' is as much a political category as it is a technical category. In some cases, villages having sources of drinking water try and get themselves declared as 'no-source' so as to access governmental schemes for rural drinking water provisioning. The 73rd Amendment to the Indian constitution gives the panchayats the power to provide drinking water in the villages. The panchayats need to be strengthened financially and institutionally so that they are able to fulfil this constitutional obligation. Most attempts at providing water for the no-source villages have generally tried to do so through tankers or pipe-water schemes. Attempts should be first made to make villages self-sufficient in drinking water by watershed treatment and similar other measures. If all measures fail then only water supply from outside the village/panchayat can be considered as the last option.

Two Issues of Contemporary Relevance

Till now inter-sectoral conflicts surrounding water usage have not been significant in India. But that does not mean there have been no important conflicts over water resources. Over the last couple of decades or so the water sector has been an important site of conflict and contestation; at the macro-level big dams and water resource development projects have been important sites of conflict between many types of social actors. Similarly, the increasingly scarce ground water resources have been important sites of contestation, albeit at a much less intense level, in many regions of the country. These need to be looked into in some detail.

The Conundrum of Ground Water

Most of the demand for both irrigation and drinking water till very recently was being met with surface water resources. But for the last two decades or so, ground water is increasingly important, and its use has been growing at a faster rate than surface water resources.

The first large-scale attempt to scientifically plan and develop ground water in India was made in 1934. The actual expansion of ground water irrigation started in 1965 after the introduction of high-yielding varieties of seeds. Increasing availability of electricity in the rural areas due to the rapidly expanding rural electrification programme, and the easier availability of credit due to the growth of the rural cooperative credit structure has helped in the setting up of a large number of wells, and in increasing the reach of ground water irrigation. Over the last few decades there have been sophisticated investigations of the ground water resources of the country, though some problems with methodology remain. The methods employed have included hydro-geological surveys, geophysical studies and exploratory drillings (Jain et al. 1977).

The rates of extraction of ground water in India far exceeds the replenishment rate in many blocks which is leading to a continuous lowering of the water table. In 2004, an alarming 28 per cent of the ground water blocks in India could be classified in the category of semi-critical, critical or overexploited; the comparable figure in 1995 stood at only 7 per cent. Most recent legislation by some state governments lack any nuanced understanding of the resource; these legislations rely mostly on controls and licenses imposed by the state, and have very little scope for community participation. As is evident from experience, both the state and the market seem to find it difficult to ensure the equitable and sustainable use of ground water. The only option left seems to be management through local community-based institutions that are adequately supported by good science. In this context enabling legal frameworks and well defined rights also assume importance (EPW 2007).

Water Resource Development and the River Interlinking Project (RIP)

As is evident from the above discussion, ground water use in India has already met its limits. Creation of surface water potential in India also seems to be plateauing. With a rapidly growing economy demand for water can only increase at a faster pace than in the past. This calls for a holistic approach towards the management of both demand and supply. But the planning for water in India has traditionally been focused on discrete, individual projects. Projects have almost always been approved on the basis of the Benefit-Cost Ratio (BCR) that is an inherently unsatisfactory criterion, and is prone to being distorted. Most of the major/medium projects are loss-making propositions for the various state governments. The cost and benefits of these projects are unevenly distributed across various social groups, and this makes rehabilitation and resettlement a huge problem, as India has one of the worst records regarding these issues. There are also major problems in terms of the planning, funding and construction of major/medium water resource development projects in the country; these include the thin spread of finances over a large number of projects resulting in cost and time over-runs, the persistent over-estimation of benefits and under-estimation of costs, ignoring of social costs and provision of inadequate funding for rehabilitation and resettlement, the persistent underestimation of negative environmental impacts, and the substantial and consistent under-utilization of the irrigation potential created (Iyer 2001a).

The proposed RIP needs to be rigorously examined keeping in mind the historical background of the experiences of large-water resource development projects. The proposal to link all the rivers in India is an old one. The present round of discussions and activities surrounding the plan have started with a writ petition filed in the Supreme Court of India following which the Court treated this issue as an independent public interest litigation (PIL) writ petition. As far as preliminary analysis shows, even if the decision were to be based on purely 'technical' matters then the RIP will lose out to other emerging technological solutions such as desalinization for meeting drinking water needs of coastal cities or improvements in the realization of existing irrigation potential for meeting additional irrigation needs. This project is purportedly an attempt at hydraulic equity at the national level that envisages transferring water from the so-called surplus areas to the so-called deficit areas. The inherent logic is flawed on many counts. First, the deficit areas are not necessarily poor or deprived; the deficit has arisen due to a process of intensive water usage in the first place. Second, the promotion of 'national' equity between regions/states would involve heightening social inequities within, between various social groups. Since there have been no professional assessment of the proposals, or even proper pre-feasibility studies, the proposal seems unacceptable on socio-economic and ecological grounds (Bandyopadhyay and Perveen 2004).

A major part of the RIP proposal envisions damming the Himalayan Rivers. There is a serious absence of long-term records of water flows, and there are practical problems in maintaining data quality leading to high levels of uncertainity in the assessment of water resource potential in these rivers (Kattelmann 1987). The so-called surpluses in the Himalayan Rivers must be properly assessed before any major river diversion schemes are planned based on insufficient data.

New Approaches and their Limitations

Due to the limitations of the water resource development paradigm mentioned in the preceding section, the last couple of decades have seen the growth of many alternative approaches. These approaches try to respond to the problems that the earlier paradigm either created or failed to address; these include piecemeal and technology-driven approaches to water resource planning, lack of focus on environmental and social concerns, and the focus on the supply rather than the demand side of management.

Watershed Development

Watershed development is an alternative way of dealing with these concerns. The last decade has seen a critique of statist resource management practices and increasing decentralization of responsibilities for natural resource management to the community level. The Watershed Development Programmes (WDPs) that are being supported by the Government of India are an example of such a change. Assessment and evaluation approaches for judging the success or otherwise of WDPs in India have evolved over a period of time. During the seventies and the early eighties, the evaluation of WDPs was premised mainly on biophysical criteria. This started changing in the late eighties with a growing recognition of the fact that watershed development involves more than just maintaining or improving the quality of natural resources, and now have objectives related to social, ecological/environmental and equity concerns as well (Turton 2000).

Increasingly, there is the diagnosis that the 'dryland blindness' of planners and policy makers has failed to recognize that climatic and livelihood uncertainty is an integral part of the lives of people in the drylands occupying much of the country. It is argued that policy makers and planners are based in distant capitals, and they are better accustomed to wetter areas with perennial rivers and more efficient irrigation facilities. These yardsticks are used to evaluate and plan water resources. This has apparently resulted in 'dryland blindness,' and has led to the spread of conditions of acute water scarcity in the dryland areas of the country. Rainwater harvesting techniques are probably the best way to tackle the water problem in drylands. The 'dryland blindness thesis' argues that water harvesting is unlikely to have the negative effects associated with large dams and schemes that tap finite ground water resources. Yet it has not gained acceptability amongst key decision makers in water resource management. The lesser visibility of watershed development schemes and their lack of appeal for powerful business lobbies may explain why watershed development has not been taken up as a political issue (Mehta 2000).

The actual experience of watershed development projects in India has been varied. In India they have had a history of at least three decades, and have been conceived basically as a strategy for livelihood protection of the people living in the fragile ecosystems that experience soil erosion and moisture stress. Under the new guidelines more than 10,000 Drought Prone Areas Programmes (DPAPs) have been started in the country over the last six years, and the overall impact of the programme seems to be positive and significant in comparison with the previous period. There has been a distinct improvement in the programme since the implementation of the new guidelines, even though it remains highly uneven across the various states (Hanumantha Rao 2000).

But such narratives of 'success' do not go uncontested. For example, a water audit conducted as part of the Karnataka Watershed Development Project (KAWD) shows that demand for water, induced by watershed 'development,' is slowly outstripping supply, and the possibility of augmentation is limited. The audit also found no evidence that watershed development activities have reversed or even halted degradation of water resources or have helped to take steps towards making villages drought-proof. Watershed development is not necessarily going to lead to either sustainable or equitable development of natural resources in the KAWAD watersheds (Batchelor *et al.* 2003). Therefore, watershed development, even in its 'participatory' incarnations, cannot be used as a general panacea for the ills plaguing the irrigation sector in India.

Traditional Water Management Systems

India has had many different kinds of traditional water management systems that evolved in response to local needs and conditions. In much of peninsular India, tanks have been important sources of irrigation. But their importance has been decreasing slowly in the post-independence period. The proportion of the gross cropped area irrigated by tanks decreased from nearly 17 per cent in 1952–53 to around 5 per cent in 1999–2000. This has to be seen in conjunction with the fact that the proportion of the area under tank irrigation has decreased more rapidly in states where tank irrigation used to be more important, and the decrease is a lot less steeper in those states where it used to be less important; in fact there have been marginal gains in tank irrigation in states that traditionally relied less on it. Data from the Agricultural Census of India across two decades from 1970–71 to 1990–99 shows that lands of small and marginal farmers owning less than 2 ha of land continue to account for the majority of tank-irrigated area in India. In 1970–71 these two categories together accounted for around 40 per cent of agricultural land irrigated by tanks whereas by 1990–91 it had increased to more than 55 per cent. During the same period of time the share of large farmers in the tank-irrigated area decreased from around 14 to 6 per cent. Thus, tank irrigation remains a crucial resource for small and marginal farmers (Palanisami 2006).

In the last decade or so there has been an increasing recognition of the importance of what are called traditional water harvesting systems that primarily consist of tanks in much of peninsular India. The work of the Delhi-based research organization Centre for Science and Environment (CSE) has been crucial in such a refocus. Tank-based systems in peninsular India have received increasing attention from both academics and practitioners. There has been an increasing recognition of the various state authorities regarding their importance, and over the last decade there have been many state-sponsored tank rehabilitation programmes. But still the actual number of tanks rehabilitated remains proportionately small compared to the total number of tanks. Augmenting tanks through peoples' participation is increasingly seen as a cost-effective as well as an equitable tool to address concerns surrounding agricultural productivity and rural poverty. Both physical and institutional measures are necessary for revitalizing tanks. In most tank revitalization programmes structural improvements tend to overshadow investments in institutional improvement. Before any investments are made in tank irrigation systems, the availability of adequate water for the tanks must be ensured. Enabling institutional pre-conditions should be created so that tank management institutions are genuinely representative and equitable (Sakthivadivel et al. 2004).

Integrated Water Resource Management

Integrated Water Resource Management (IWRM as it is more popularly known) is a part of a host of new strategies being promoted by international bodies as innovative practices in the water sector. Indian water policy discussions are generally influenced by overarching global discourses. The discourse surrounding IWRM reflects an emerging global consensus surrounding water; the main elements of this consensus are a naturalized and absolutist notion of water scarcity and its links to poverty and deprivation, and a host of demand-management practices and policies. There are serious flaws in such a diagnosis and prescription; the link between water scarcity and 'water poverty' is tenuous. The main components of IWRM packages, such as reforms in water pricing for water to reflect its 'production costs,' the enforcement of water withdrawal permits have generally been difficult to implement. Attempts at 'integrated planning' at the regional level have failed in a similar manner. One of the biggest reasons for such failures has been that water economies in the poor countries are more often than not informal in nature, and this makes the operation of formal demand management systems that are imposed from above almost impossible to succeed. But IWRM cannot be completely rejected. It has its applications in highly formalized segments of the water economy. With increasing urbanization, our cities with a significant part of their water economy in the formal sector will need the direct demand-management practices as envisaged by IWRM (Shah and van Koppen 2006).

Even in the rural sector, with increasing water scarcity, well-regulated water markets seem to have emerged in certain regions. In these cases also IWRM can be a fruitful framework to locate interventions for producing desirable outcomes with respect to both equity and efficiency. As a framework, IWRM cannot be rejected summarily without figuring out its applicability in specific sectors and regions. Till very recently, the water sector has been dominated by supply-side concerns, and by the various disciplines of engineering. IWRM provides us with a framework to integrate social science concerns with engineering know-how, and to address demand-side management as well. This is a welcome development, and IWRM should be given a fair, if a limited trial as a framework for locating water-related interventions in the country.

Water Markets and Appropriate Pricing of Water

Due to the apparent inefficiencies in state-managed irrigation systems, market-based mechanisms are increasingly being promoted as alternatives. There are two broad sets of reasons that are given for preferring water markets over administered pricing. Theorists advocating market mechanisms for allocating water resources generally claim that it might be a theoretical possibility to devise and implement an efficiently administered price system, but in the real world the information requirements are demanding, and can hinder the effective functioning of such a system. An administrative solution presumes neutral, efficient and incorruptible administrators and administrative bodies that design and implement the 'correct' prices. More often than not in practice, this is not borne out; these administrative bodies are more often than not captured by interest groups or are ill-informed about future demand or are inefficient and are unable to set and send out the correct price signals or collect the water charges effectively (Mohanty and Gupta 2002). These are, in short, some of the main arguments given for water markets.

Critiques of free-market solutions argue that in actual experience, water markets need regulation and the enforcement of private rights, and they run into the same set of problems as administered pricing. More importantly, there are many examples of market failures, and the introduction of water markets is no guarantor of an efficient mechanism for allocating water. The informal water economy of countries like India involves complex intertwined claims over water, and market mechanism cannot be simply grafted over these. Questions surrounding water equity also assume salience in the context of water markets. In this context communitarians argue in favour of a third way that steers away from both statist and market mechanisms by giving effective rights to communities to manage their own water resources (Iyer 2001b).

But in the case of water, there are other possible responses to the slogan 'get the price right'. In a classic case of correct diagnosis being followed by incorrect prescription, it does not necessarily follow that the only, and the most effective, way to conserve water and increase irrigation efficiency is to increase prices in order to reflect the value of water scarcity. As suggested by some scholars there are two broad reasons for reaching this conclusion. First of all, raising the prices of canal water to the point where it significantly affects water demand will have a negative impact on farm revenues in the short and medium term, and moreover such a policy will be politically indefensible. And low-price increases will have very little impact on water demand. More importantly, the choice of water-inefficient crops cannot be squarely laid at the door of low water prices of canal water. Farm-level inefficiencies are not the most significant on existing systems of canals; the inefficiencies instead lie at the higher levels of the system. A better mechanism can be to enforce simple allocation rules (such as a per-hectare ration) that can immediately signal the value of water scarcity without raising prices. This is politically a far more feasible step than raising prices of canal water significantly, as quantitative restrictions over water use are already a part of the management of most Indian canal systems. A transparent and equitable quota system can also free up water to be transferred to urban usage, or to other farmers, or to meet environmental needs. This is not to argue that one should not 'get the price right;' unless and until mechanisms for effective price signals are in place. 'Getting the price right' is not automatically the best possible way under most circumstances in India to deal with the inefficiencies in irrigation water allocation (Ray 2005).

Conflicts, Constitutional Provisions and Institutions

Apart from 'getting the price right,' 'water wars' is another strand around which contemporary narratives surrounding water are being framed. Despite the increasingly audible international rhetoric surrounding 'water wars,' most incidence of violence surrounding water is at the sub-national level. These conflicts can be between sectors, as is the case with the farmers' protests surrounding the transfer of water to non-agricultural use from the Hirakud reservoir in the Indian state of Orissa, to those between different states, as is the case of the dispute surrounding the waters of the Cauvery River between the states of Karnataka and Tamil Nadu, or those that involve governments and non-governmental actors, as is the case with the struggle of the Narmada Bachao Andolan (NBA) that has been fighting against the Sardar Sarovar Project (SSP) on the Narmada River with the focus being on the problems of resettlement and rehabilitation. Most of the conflicts seem to be clustered around large projects and on inter-state rivers.

Conflict Resolution

With regards to inter-state river disputes, Article 262 of the Constitution and the Inter-State Water Disputes Act (ISWD Act) 1956 provides the adjudication system as a mechanism for conflict resolution. This Act needs to be properly implemented by reducing delays at every stage. In addition to making the adjudication process better, it is equally important to provide an enabling environment for conflict resolution through negotiation. Institutional means should be provided for achieving this, and River Basin Organizations (RBOs) can perhaps fulfil this role. Conflicts between states are only one of the many types of conflicts that arise surrounding water, and there needs to be an enabling legal, institutional and policy environment to help deal with all these various types of conflicts effectively; such mechanisms are sadly lacking now. In a policy environment increasingly suffused by neo-liberal prescriptions, a system of clearly defined water rights is sometimes proposed to help reduce and resolve water conflicts. This is a simplistic view as this prescription does not take into account the already existing set of claims and rights over water; in fact the very attempts at enforcing such rights might lead to further conflicts. The set of conflicts between the people and the state needs to be looked at a lot more carefully (Iyer 2001b).

Constitutional Mechanisms

The most important entry in the Indian constitution on water is Entry 17 in the State List that deals with water supplies, irrigation, canals, embankments, water storage and hydro-power. This entry is subject to the provisions of Entry 56 of the Union List of the Constitution that primarily deals with inter-state rivers, and gives sweeping powers to the central government to regulate and develop inter-state rivers in 'public interest'. But the centre has not been able to make any substantial use of the enabling provisions of this entry. The 11th and 12th Schedules to the Constitution list drinking water, water management, watershed development and sanitation to be devolved to the panchayats and urban municipal bodies. These bodies will have an important role to play in the future, and future legislation will need to take their existence into account (Iyer 2003).

Institutional Reform

The National Water Resources Council (NWRC) set up by the Government of India, although an important body in the Indian water sector, has no statutory backing and therefore lacks teeth. Whether giving the NWRC statutory backing will improve matters of federal water governance is a difficult question to answer. One long-standing institutional innovation that is yet to be implemented is that of the RBOs. Despite the hostility of the states to River Basin Organizations this needs to be done on an urgent basis as water management at the basin-level as a whole can go a long way in diffusing tensions between various stakeholders and in reducing the intensity of the conflicts. The Inter-State Council, a recommendory body established by Presidential Order on 28 May 1990, is another institution which can be utilized to settle disputes surrounding the waters of inter-state rivers (Iyer 2003).

Water and Equity

As the preceding short discussion shows, the Indian constitution primarily frames the debates surrounding water through spatial metaphors and statist provisions. The provisions of the constitution look at potential conflicts primarily as between the various actors of the state. But the neglect of the social aspects of water usage within the Indian constitution is not unique. Even academic work on social issues surrounding water have, more often than not, focused on issues of efficiency; i.e., primarily around issues of cost recovery of irrigation water and the social reasons for non-sustainability of water usage. Issues surrounding equity have started receiving some attention of late. The discussion that follows focuses on only two axes surrounding which issues of equity can be raised; the first one being that of gender, and the second one being that of ethnicity, for instance of tribals in India. There are many other axes on which such analyses can take place — this discussion is merely indicative of the potential of such an exercise.

Women and Water

In international public-policy concerns surrounding water women rarely figured in the fifties and sixties. By the seventies and eighties, it came to be recognized that women had an important role to play in domestic water provisioning. They were identified as important users of water as a resource and their role in the broader context of managing water as an environmental resource also started to be recognized. During the eighties (the UN Development Decade that focused on water), the focus on women in the context of water management was mostly to make the systems more efficient economically, and did not overtly factor in gender-related concerns. In the nineties with the Washington Consensus firmly in place, water came to be increasingly seen as an economic good, and women's roles as water managers came to be foregrounded. More recently it's the discourse on poverty alleviation and empowerment that frames discussions surrounding gender and water (Wallace and Coles 2005).

In the context of these changing policy orientations, it must be affirmed that in most poor rural and urban communities it is the women and the girls who have the responsibility of collecting water for domestic needs. Water collection chores do not affect all women equally as they are not one homogenous group; the socio-economic status of the household, age and marital status, seclusion and household composition affect women's water-related activities. Till very recently government departments and NGOs focused on the provision of drinking water, and for all practical purposes ignored the issues of sanitation and waste-water disposal. Lack of accessible sanitation has significant impact on women: women bear the cost of lack of proper sanitation facilities more than men do, and they have relatively less command over financial resources to make their needs for sanitation felt in households and communities. The importance of women to the water sector and vice versa is not limited to domestic water and sanitation alone. In India, women work as agricultural labourers and farmers (sometimes as legal owners of farmland, or mostly as de facto household heads). Women are the invisible backbone of Indian agriculture, and a significant proportion of the workforce engaged in agricultural labour comprises of women. Even when women are actively engaged in managing farms, they are generally not recognized as 'farmers'. More often than not their water rights are dependent on male land-owners as water rights in India are tied to land, which in this country are primarily under the control of men. There is an urgent need to articulate women's rights to land and water through legal and policy changes (Kullkarni *et al.* 2007).

It must be evident from the above discussion that water use amongst women happens along multiple axes and they play an important role in using this resource in the sustenance of the household economy. If this role of women is recognized it can go a long way in promoting a more efficient and productive use of water in strengthening rural livelihoods. Studies have shown how easier access to water for women in production can help to strengthen the family livelihood by increasing their incomes; it can also aid in gender empowerment in a significant manner by strengthening the bargaining position of women in a significant manner (Upadhyay 2005).

The Scheduled Tribes and Water

Various surveys conducted in the tribal areas of peninsular India reveal that tribal farmers earn much less income from their land compared to the non-tribals in tribal-dominated districts. Both the investments and the returns on the investment made by the tribals are lower compared to the non-tribals in the same area. Water resource development in India has traditionally been a technology-driven, statist project, and has not been suitable for the tribal people as they usually live and work in the uplands. The explosive growth of ground water irrigation in recent times has also bypassed them. In planning water-related interventions in tribal areas there is a need to be sensitive to factors such as agro-climatic conditions, local topography, level of local infrastructure development and the socio-cultural characteristics of the communities. Water-related interventions should also have a labour component involving the 'Food for Work' mechanism. Carefully implemented and sustainable water-related initiatives in these areas can help in reducing the deprivation amongst tribals and can help in enhancing national food security. This can also contribute towards sustainable and more efficient utilization of our land and water resources (Phansalkar and Verma 2004).

It also has to be kept in mind that not only have the benefits of traditional water resource development projects bypassed tribals; it is they who have borne most of the social costs of these projects. Therefore, any new water resource development projects should pass scrutiny in terms of benchmarks of ensuring the rights of tribals over land and water.

Conclusion

India lacks suitable legal frameworks and institutional mechanisms to manage the water sector efficiently and equitably. Irrigation is the biggest consumer of water and is the reason for most of the conflicts surrounding it. One of the major issues in irrigation is the low efficiency of water usage that stands at around 30–40 per cent in the case of canal irrigation systems. This needs to be urgently addressed. The conventional approach has been to lay the blame squarely on farmers, and the low cost of irrigation water that apparently also leads to wastage. A big part of the wastage in irrigation water happens at the supra-farm level; increasing prices will not help us to deal with this. This is not to say that pricing is not an issue; it is. But pricing reforms can only be part of a larger set of reforms that address broader structural and institutional concerns.

Benefits from major canal irrigation systems and most of the subsidies on irrigation water tend to go to the landed sections of rural areas. Major policy changes are needed to put social equity in the irrigation sector at the centre of our concerns. One of the initiatives in such a context can be to ensure that all the farmers growing the same crop should have an equal share in both opportunity and costs. In practical terms this means that if a farmer grows rice he must pay the same amount for the water he receives regardless of whether he is a part of a canal system or is withdrawing water from the underground aquifers. Another mechanism can be to share benefits in the sense of exchangeable rights over water rather than actual water per se. Ground water irrigation can be made more sustainable by moving from a system of free/cheap electricity for farmers to a system of minimum and exchangeable electricity quotas for every inhabitant of rural areas beyond which marker rates will be charged.

Full-cost recovery at the unit level in all urban centres drawing water from beyond a certain specified radius (say 5–10 kms) should be made compulsory. Water charges can be made directly proportional to the property taxes paid, and the poor can be cross-subsidized at the cost of the bigger and richer users of water in the urban areas (Gujja and Shaik 2005).

The discourse on water is now dominated by the water resource development paradigm that essentially tries to meet future demand projections through big projects aimed at storing or transferring water. As already detailed, there are many problems with such an approach. As our discussion of the newly emerging approaches to water management shows, one cannot whole-heartedly embrace the 'alternatives' also. The water sector is now dominated by dichotomous arguments; debates between state vs. market, market vs. communities, prices vs. quota and small vs. large systems. Such a dichotomous approach needs to change; the need of the hour is context-specific thinking that takes into account all the factors in a given situation.

Large-scale storage projects are, more often than not, a bad idea, especially if they are not fully integrated into extant small systems starting from the planning stage itself. Ecological sustainability should be a guiding factor in all our considerations of water planning; water is not a resource such as, say, coal or oil; it is vital for the existence of life itself. Solutions on both the supply- and demand-side are needed, and the right mix has to depend upon the context; but, more often than not demand-side management seems to be a better option than supply-side interventions. Recognition of community rights of usage is necessary to aid efficient and equitable water management practices.

Although efficiency, both technical and economic, has to be addressed as a major issue of public policy surrounding water, equity has to be at the centre of all water-related interventions. There is a strong case to be made for dissociating water rights and land rights; the landless must receive proportionately larger share of rights over water to offset existing inequities in terms of land distribution. The powers of the gram sabha over land and water resources in the Fifth-Schedule areas should be enhanced, and their already existing powers should be respected. This can go some way in ensuring that the state does not violate the rights of the scheduled tribes in these areas. Water is essential to life, and this makes water equity central to all discourses surrounding equity in India. Perhaps one should mention here that equity and justice are not only matters of policy prescriptions, but also matters of struggle. In this context, popular struggles matter as much as policy and legal changes; thus conflicts need not always be problematized.

Note

1. For a slightly more detailed discussion of water availability in India, please refer to Chapter 4 of this volume.

References

Bandyopadhyay, J. and Shama Perveen. 2004. 'Interlinking of Rivers in India: Assessing the Justifications', *Economic and Political Weekly*, 39 (50): 5307–16.

Batchelor, C., M.R. Rama and M. Rao. 2003. 'Watershed Development: A Solution to Water Shortages in Semi-Arid India or Part of the Problem?', *Land Use and Water Resources Research*, 3: 1–10.

Bhaduri, A. and Arvind Kejriwal. 2005. 'Urban Water Supply: Reforming the Reformers', *Economic and Political Weekly*, 40 (53): 5543–45.

EPW Editorial. 2007. 'Ground water: Half-Solutions to Ground water Depletion', *Economic and Political Weekly*, 42 (40): 4019–20.

Gujja, Biksham and Hajara Shaik. 2005. 'A Decade for Action: Water for Life When Will India Cover the "Uncovered"?', *Economic and Political Weekly*, 40 (12): 1086–89.

Hanumantha Rao, C.H. 2000. 'Watershed Development in India: Recent Experience and Emerging Issues', *Economic and Political Weekly*, 35 (45): 3942–47.

———. 2002. 'Sustainable Use of Water for Irrigation in Indian Agriculture', *Economic and Political Weekly*, 37 (18): 1742–45.

Iyer, R. R. 2001a. 'Water: Charting a Course for the Future: Part I', *Economic and Political Weekly*, 36 (13): 1115–22.

———. 2001b. 'Water: Charting a Course for the Future—II', *Economic and Political Weekly*, 36 (14 &15): 1235–45.

———. 2003. *Water: Perspectives, Issues, Concerns*. New Delhi, Thousand Oaks and London: Sage Publications.

Jain, B. K., B. H. Farmer, H. Rush, H. W. West, J. A. Allan, B. Dasgupta and W. H. Boon. 1977. 'India: Underground Water Resources [and Discussion]', *Philosophical Transactions of the Royal Society of London. Series B, Biological Sciences*, 278 (962): 507–24.

Janakarajan, S. and Marcus Moench. 2006. 'Are Wells a Potential Threat to Farmers' Well-being? Case of Deteriorating Ground Water Irrigation in Tamil Nadu', *Economic and Political Weekly* 41 (37): 3977–87.

Kattelmann, Richard. 1987. 'Uncertainty in Assessing Himalayan Water Resources', *Mountain Research and Development*, 7 (3): 279–86.

Kulkarni, Seema, Sara Ahmed, Swarnalata Arya, K. J. Joy and Suhas Paranjape. 2007. *Women, Water and Livelihoods*. Pune: Women and Water Network and Society for Promoting Participative Ecosystem Management (SOPECOM).

Mehta, Lyla. 2000. 'Drought Diagnosis: Dryland Blindness of Planners', *Economic and Political Weekly*, 35 (27): 2439–45.

Mohanty, N. and Shreekant Gupta. 2002. *Breaking the Gridlock in Water Reforms through Water Markets: International Experience and Implementation Issues for India*, Working Paper Series, Julian L. Simon Centre for Policy Research. Liberty Institute. Available at (http://www.libertyindia.org/policy_reports/water_markets_2002.pdf.

Palanisami, K. 2006. 'Sustainable Management of Tank Irrigation Systems in India', *Journal of Developments in Sustainable Agriculture*, 1(1): 34–40.

Phansalkar, S. J. and Shilpa Verma. 2004. 'Water and Livelihoods: Improved Water Control as Strategy for Enhancing Tribal Livelihoods', *Economic and Political Weekly*, 39 (31): 3469–76.

Ramachandraiah, C. 2001. 'Drinking Water as a Fundamental Right', *Economic and Political Weekly*, 36 (8): 619–21.
Ray, Isha. 2005. '"Get the Price Right": Water Prices and Irrigation Efficiency', *Economic and Political Weekly*, 40 (33): 3659–68.
Sakthivadivel, R., P. S. Gomathinayagam and Tushaar Shah. 2004. 'Rejuvenating Irrigation Tanks through Local Institutions', *Economic and Political Weekly*, 39 (31): 3521–26.
Shah, Tushaar. 2004. 'Water and Welfare: Critical Issues in India's Water Future', *Economic and Political Weekly*, 39 (12): 1211–13.
Shah, Tushaar and Barbara van Koppen. 2006. 'Is India Ripe for Integrated Water Resources Management? Fitting Water Policy to National Development Context', *Economic and Political Weekly*, 41 (31): 3413–21.
Tejwani, K. G. 1987. 'The Himalaya–Ganges Problem', *Mountain Research and Development*, 7 (3): 323–27.
Timmons, John F. 1954. 'Economic Framework for Watershed Development', *Journal of Farm Economics*, 36: 1170–83.
Turton, Cathryn. 2000. *Enhancing Livelihoods through Participatory Watershed Development in India*. Working Paper 131. Overseas Development Institute.
Upadhyay, Bhawana. 2005. 'Gendered Livelihoods and Multiple Water Use in North Gujarat', *Agriculture and Human Values*, 22: 411–20.
Wallace, Tina and Anne Coles. 2005. 'Gender, Water and Development: An Introduction', in Anne Coles and Tina Wallace (eds), *Gender, Water and Development.* Oxford and New York: Berg.

Section II
Institutional Framework

3
Boundaries of Transboundary Water Sharings

Narendar Pani

Introduction

Inter-state river water disputes in India have a tendency to be intractable, and the dispute over sharing Cauvery waters is far from being an exception. The current phase of the dispute, which can be said to have begun when Tamil Nadu sought adjudication in 1969, has shown all signs of continuing even after the Cauvery Water Disputes Tribunal came up with its final order 38 years later. In the search for a solution the states in the dispute have left virtually no detail untouched. The documents and publications brought on record before the Tribunal by the parties concerned in support of their respective claims run into more than 50,000 pages (CWDT 2007).

This concern for detail is neither unusual nor avoidable. Each river, and its basin, has its own unique history. This uniqueness is reflected in the role the river plays in the culture of the basin. The human element in the basin too is rarely constant. The stakeholders in the river can keep changing over time, as a result of the social, economic and political changes in the basin. And even when the same group of individuals retains a stake in the river, the stake itself need not be the same. Lifestyles of some groups can change, making them less dependent on the river, even as the lack of options may increase the dependence of other stakeholders on river waters. It does not require great insight to recognize that any dispute settlement mechanism can hope to be effective only if it is extremely sensitive to detail. And yet there is the possibility of getting lost in the detail. The detail has a way of engulfing us; of narrowing our vision to a set of issues that attract the greatest debate. And since the issues that attract the most intense and long drawn out debates are, almost by definition, the ones that are the most difficult to agree upon, the dispute itself becomes even more intractable than it already is.

When the details are so daunting it may be worthwhile to step back and take a larger view. A broader canvas could help bring in dimensions that offer more options in the negotiating process. As the potential benefits of cooperation between states increases, they would be less inclined to allow water sharing disputes to prevent cooperation in other areas. And the benefit from a larger canvas need not be confined to areas of cooperation alone. There may also be negotiating benefits from linking disputes. A state could be prompted to offer greater concessions in water sharing if it can get concessions in some other area in return. This does not, however, mean that any expansion of the list of issues will always be useful. On the contrary, an indiscriminate linking of issues could easily turn out to be counter-productive. Indeed, the Cauvery dispute is itself a striking example of what can go wrong when water sharing is linked to a more emotional dispute such as one on language. The conflict between Tamil and Kannada has had a long history within Karnataka, particularly its capital city, Bangalore. Bangalore, had till independence in 1947, two distinct cities, the 'city' and the 'cantonment'. The cantonment was directly under the rule of the British government in what was then the Madras Presidency (Pani *et al*. 1985), and its most extensively used language was Tamil. With independence the cantonment was integrated with the city, but the social isolation between the two areas remained for decades after. And when the social integration finally gathered momentum in the 1980s, it was accompanied by a strong Kannada movement in what is now known as the Gokak agitation (Nair 2005). And the conflict was not prompted by the city-cantonment divide alone. Even within the city, the ethnic distinction between Kannadigas and Tamils was not always easy to ignore. When the British returned the state of Mysore (and Bangalore City) to indirect rule in 1881 they simultaneously introduced a Mysore Civil Service that was originally dominated by officers trained in what is now Tamil Nadu. When this dominance was challenged, successfully, within the bureaucracy in Mysore it was not without its own ethnic tensions. These ethnic and language disputes were allowed to get closely enmeshed into the water sharing disputes since the 1990s, adding an emotional edge to a water sharing dispute that was already difficult to solve.

The widening of the canvas cannot then be an indiscriminate process. It would be necessary to recognize that there are issues that form an intrinsic part of any water sharing dispute and those that are not. In other words, even as we look beyond water sharing alone it is important to be aware that the exercise cannot be in operation without fresh boundaries that separate the issues that are relevant to river water disputes from those that are not. In deciding what are intrinsic, two temptations will have to be resisted. The first is the temptation to bring in extraneous considerations only because it is believed they will help ease pressures. Discussions around inter-state river water disputes in India sometimes invoke national interest, or the we-are-Indians-first argument, to help resolve the dispute. But since the Indian identity has little, if any, bearing on the issues the river has thrown up, the shared identity of all the parties in the dispute generally tends to be ignored. The second, and corresponding, temptation is to leave out issues that are difficult to resolve. It may even be possible to arrive at an agreement on some issues, leaving out the ones over which deep divisions remain. Such an approach may have short-term value as it could help ease tensions over a dispute. But if the issues that have a bearing on the river are left unaddressed for long they have the potential to turn into a festering wound. In deciding on the issues that need to be considered then we need to be comprehensive in choosing the issues that have a bearing on the river, even as extraneous considerations are kept out.

This approach clearly places a premium on our understanding of what has a direct bearing on water disputes and what does not. In arriving at such an understanding we would have to take into account the experiences of river water disputes across the world. The diversity of these experiences is an in-built protection against a mechanical transfer of dispute resolution mechanisms from one basin to another. At the same time, the common features that can be seen despite this diversity would be useful pointers to the nature of river water disputes and their resolution. These pointers could help in putting together a framework for understanding the dynamics of river water disputes. This framework could then be adapted to deal with the specifics of each dispute. While this exercise would be particularly rewarding in cases, like the Cauvery, where the arguments find it difficult to emerge from the detail, our purpose in this paper is confined to identifying a framework for deciding the boundaries of transboundary water disputes.

Scarcity and Conflicts

Conflicts over river waters are often linked directly, and not always justifiably, to the issue of scarcity alone. This is not merely because each party to a dispute needs to cite scarcity to make its claims for more water credible. But these particular arguments are often taken to a general conclusion that conflicts over river waters are the result primarily, if not solely, of water scarcity. It is this thinking that underlies the increasingly popular belief that the growing scarcity will inevitably lead to water wars.

In an obvious sense, scarcity of any resource, let alone one as essential as water, would tend to intensify any conflicts that already exist among the different users of that resource. And there is little doubt that the numbers of those across the world facing acute scarcity of water is growing. Hydrologists conventionally use 1700 cubic metres per person as the national threshold for meeting water requirements for agriculture, industry, energy and the environment; 1000 cubic metres is taken to indicate water scarcity and less than 500 cubic metres, absolute scarcity. By this indicator 700 million people in 43 countries live below the 1700 cubic metres per person mark, with Palestinians experiencing the greatest water stress at 320 cubic metres per person (HDR 2006). And this indicator may even be a gross underestimate as it goes by the national average. By ignoring distribution it does not quite take into account the water scarcity in specific regions within countries where the overall average may be over the stress level. This problem is particularly true of large countries, where the area covered is so great that they can easily cover both water scarce and water surplus regions. The fact that the average availability in the two sets of regions taken together is above the threshold level is of no help to the water scarce regions. This factor is particularly severe in countries like India. It has been estimated that over 224 million people in India live in river basins where the renewable water resources are below the 1000 cubic metre water-scarcity level (HDR 2006).

At the core of this growing scarcity is the phenomenal increase in water withdrawals. Much of the growing crisis in water is because of over-utilization of this resource. Water withdrawals across the world have increased from 500 cubic kilometres in 1900 to 3830 cubic kilometres in 2000 (HDR 2006). And the economic factors that have caused this massive increase in withdrawals show no signs of abating. A major contributor to this increase has been agriculture. In countries that are short of food security, there is an understandable reluctance to limit the amount of water supplied to agriculture. In theory, economists could argue that the food deficit countries could import from regions where food is produced with a more efficient use of water. In reality though, it is difficult, not without reason, to get countries to rely entirely on the international market for food security. Apart from the volatility of prices in that market, it also has implications for international politics. It is not surprising then that most countries would like to ensure food security primarily through domestic production. And if this requires the supply of highly subsidised water for agriculture, so be it.

Meeting this requirement on a sustained basis is also not always a matter of the overall availability of river water resources. Even when the water available is sufficient to produce the required food, it could get diverted to other ends. Within agriculture itself the water resources available for foodgrains could be reduced by the water being diverted to other crops. India is an emerging example of this trend. The Green Revolution beginning in the mid-sixties ensured that India domestically produced the foodgrains it needed. But as agriculture, particularly after the mid-eighties failed to keep pace with the overall growth in a liberalizing economy, foodgrain production did not always provide the economic returns that the rest of the economy was beginning to enjoy. Farmers could hardly be faulted then for moving to some more lucrative non-food crops, even if they were more water intensive.

The dependence on river waters is also likely to grow due to the failure of other sources. Large areas in south India are dotted with tanks. These tanks, which are often centuries old, once met both the irrigation and domestic water requirements of a large number of villages. Traditionally, these tanks were desilted, and otherwise maintained, as the farmers required the silt for cultivation. With the coming of the Green Revolution and its chemical fertilizers the farmers no longer felt a need for silt. The silt was allowed to gather in the tanks, dramatically reducing their capacity in terms of the water they held and the area they could irrigate. Silt also reduced the efficiency of the tanks in replenishing ground water sources. The shift in demand from tank water to river waters is, if anything, even more acute in urban areas. Rapid urbanization has seen a number of tanks becoming history. Too often these tanks are allowed to dry, so that they can be converted entirely, or at least partially, into valuable real estate. Add to this the growing industrial demand for water, as well as the wastage, and the urban demand for river waters too has been growing. This is most evident in the increasing dependence of Bangalore on the Cauvery to meet the city's drinking water needs. There is then little doubt that water scarcity is growing at an alarming rate, increasing the pressure on river waters, and this is bound to accentuate river water disputes that already exist.

Serious as the crisis of water scarcity may be, it is important to stay clear of the popular temptation to see river water conflicts as a matter of scarcity alone. Such a view would lead us inevitably to the conclusion that increasing scarcity will result in water wars. And as has been pointed out, this has not quite happened, though Anwar Sadat suggested in 1979 that water was the only issue that would prompt Egypt to declare war again. But the absence of historical evidence is not sufficient reason to take any position in as volatile an area as water. In a dynamic, rapidly changing field like water conflicts, it would be foolhardy to assume that there will never be a first time. Our case against treating river water conflicts as entirely a matter of scarcity alone would have to be based on other factors as well. And there are at least two other trends that substantiate our case.

First, if scarcity were the only reason for river water disputes, the regions of the greatest scarcity would also be the ones with the most intense water conflicts. But this is not the case. On the contrary, some of the more lasting water-sharing arrangements have been in areas of intense water scarcity. Wolf (2000) identifies the Berbers of the High Atlas Mountains and the Bedouin of the Negev Desert as two communities that have worked out systems of water sharing in drylands. Within India too the really water-scarce regions of the Thar Desert are generally not at the forefront of the more volatile water conflicts. Indeed, it is rivers like the Cauvery or those in Punjab and Haryana, with a much higher availability of water, that have tended to be more prone to conflict, at least in recent years.

Second, and rather more important in some current contexts, measures to reduce scarcity may in fact create new conflicts. This becomes quite evident when we look at the role of distribution in the problem of water scarcity. Taken as a whole, the world has enough water to avoid stress. The problem of scarcity arises because it is not evenly distributed. Nearly a quarter of the world's supply of fresh water is in Lake Baikal in Siberia. The availability of water in one part of the world is often of little use to those starved of water on the other side of the globe. The idea of transferring water from the surplus regions to the deficit ones thus does not find too prominent a place in international literature on water. It is only where the surplus and deficit regions are within the same country that it becomes much more tempting to consider such transfers. This is particularly true of India where over two-thirds of its renewable water resources serves a third of the population. There has been intermittent official support for the idea of linking rivers to bring about such a transfer. But it is important to remember that the interlinking of rivers across states will inevitably raise the question of the rights of each state to this water. And states can be expected to have conflicting views on which particular areas should get the benefit of additional irrigation water. Indeed, just as scarcity can act as a trigger to conflict, so can the conflicting aspirations among different regions to make the transition from the economic pressures of dryland agriculture to the prosperity of irrigation. Conflicts over water can arise not just when the overall availability reaches a point of acute distress, but also when the demand for water grows to a point that makes existing supplies inadequate. And the promise of large-scale transfers of water is bound to boost demand. What is more, such a large project will also displace a large number of people. And as the Narmada dam issue has shown, conflicts over displacement can be as intense as conflicts over water sharing.

Thus, while scarcity undoubtedly has the potential to accentuate conflicts over river waters, it would be a mistake to treat it as the sole cause for conflicts. On the contrary, there can even be situations where efforts to reduce scarcity may only raise fresh conflicts. Conflicts over river waters then need to be understood separately, and not as mere by-products of the crisis of water scarcity.

Conflicts and Institutions

Conflicts over water may be caused by a variety of interests ranging from the economic and political to the legal. The specifics of a dispute could vary from sharing irrigation water, to effects of hydroelectric power generation, pollution or even something as apparently benign as fishing, if it reduces the fish stock downstream. Conflicts can range from disputes between individual farmers, to differences between groups of farmers whether they are at the level of the village, the taluk, the district, the state or the nation. As Aaron Wolf puts it, 'Transboundary water disputes occur whenever demand for water is shared by any sets of interests, be they political, economic, environmental, or legal. Conflicts over shared water resources occur at multiple scales, from sets of individual irrigators, to urban versus rural uses, to nations that straddle international waterways' (Wolf 2004: 1).

These rather wide range of issues and interests that are involved in water conflicts would make common patterns the exception rather than the norm. Indeed, some of the broad patterns that emerge from this large canvas may themselves only point to fresh diversity. For instance, the degree of violence may vary depending on the scale of the dispute. While water wars may not, fortunately, have become a reality, violence over more local disputes, including the one over Cauvery waters, is not unknown. As Wolf, Kramer, Carius, and Dabelko have pointed out, 'Some research even suggests that as the geographic scale drops, the likelihood and intensity of violence increases.' (Wolf *et al.* 2005: 87). To seek out a common pattern in this diversity would require painstaking effort. But the reward would be in the fact that any common pattern that emerges from such diversity would be a significant contribution to an understanding of river water disputes.

It is in this context that the efforts of Aaron Wolf, Shira Yoffe and Mark Giordano in looking at conflict and cooperation over water in the second-half of the 20th century gains significance (Wolf *et al.* 2003). Their study covered 1831 events covering both conflicting and cooperative events, with cooperative events being more than twice as common as conflictive events. On the basis of this extensive survey the study came to the conclusion that:

> In general, we found that most of the parameters commonly identified as indicators of water conflict are actually only weakly linked to dispute. These parameters include: climate, water stress, population; dependence on hydropower, dams or development per se, level of development or 'creeping changes,' such as gradual degradation of water quality or climate-change-induced hydrologic variability. In fact, our study suggests that institutional capacity within a basin, whether defined as water management bodies or treaties, or generally positive international relations are as important, if not more so, than the physical aspects of a system. The relationship was hypothesized as follows: 'The likelihood and intensity of dispute rises as the rate of change within a basin exceeds the institutional capacity to absorb that change.' (Wolf *et al.* 2003: 51).

The main strength of this result, which is perhaps not adequately emphasied by the authors, is the role it allows for the situational dimension in river water conflicts. Once we acknowledge that the commonly cited parameters, including such apparently obvious ones as water stress, are only weakly linked to water conflicts, we are in effect recognizing that water conflicts cannot be reduced to a set of major causes. Instead, we need to focus on the changes that give rise to disputes and the institutional ability to manage that change. The main strength of the formulation by Wolf, Yoffe and Giordano is that it can, potentially, cover virtually any situation that arises in a river basin. And yet it may not be comprehensive enough to cover everything that can affect and be affected by river waters. This formulation focuses solely on what happens within a basin. But with economic integration, conditions affecting people within a basin, and therefore their dependence on the river itself, can be quite closely linked to what happens outside it. The availability of alternative sources of electric power outside the basin, for instance, can reduce the pressure to move into those hydroelectric projects where the costs of displacement are very high. Again, the availability of employment outside the basin could reduce the numbers of those dependent on the river for their livelihood. We may then be better served by taking a broader view than that taken by Wolf *et al.* and not confine our analysis to the basin. The potential for conflict or cooperation centred around a river depends on the institutions being in tune with the changes taking place that directly or indirectly affect all those who have a stake in the river basin.

Change as Conflict

The broadening of our focus to cover all those directly or indirectly affected by a river automatically increases the scope for change. But even if we, just for a moment, confine ourselves to those who have a direct stake in the river, the potential for change is still very substantial. Political and economic relationships are often continually evolving. Even the most apparently stable of these relationships could have elements of instability built into them. For instance, a stable political equation may ensure that a relationship between various stakeholders is accepted for decades. But a perceived inequality within those relationships could act as a festering wound. And over time these wounds get difficult to ignore and tend to have an impact on the overall equations between stakeholders. There is perhaps no better example of the inequalities forcing change even in a relatively unchanging political context than the case of the water rights of the American Indian tribes. The conflict between American state governments and American Indian tribes has been termed 'one of the most divisive intergovernmental conflicts in the history of the United States' (McCool 1993: 85). The conflict between the tribes and the others in the states has been extremely intense. Of the many dimensions of this conflict, water has been a critical one. For over a century the federal government has been the official trustee of the tribes, but it has been regarded as supporting the states and other private interests. This has led to hundreds of court battles over the years. This litigation had little effect in terms of resolving the disputes. What it did instead was create an uncertainty about existing systems of water rights. It is this uncertainty that led federal governments, from the Carter administration onwards, to look at negotiations as a means of resolving these disputes. Thus though the tribes never really meaningfully challenged the political authority of either the states or the federal government the sheer inequality in the relationships created sufficient uncertainty to force a change in the institutions that dealt with water rights.

At other times the change may not even be so much in terms of tangible factors like water sharing as it is on relatively more intangible concerns as security. The most obvious case of a dispute that has been accentuated by changed security perceptions is the Indus Water Treaty of 1960. Like any other agreement between India and Pakistan, the Indus Water Treaty did not have the benefit of a politically supportive climate for its operation. Yet the treaty did survive two wars between the two countries, in 1965 and 1971 primarily because its water sharing mechanism was based on distribution of rivers rather than the water in them. The Indus Water Treaty 'accomplished this by getting India and Pakistan to consent to the permanent partitioning of the Indus River system — India winning unfettered ownership of the waters of the three eastern rivers (Ravi, Beas, Sutlej), and Pakistan acquiring nearly unfettered ownership of the waters of the three western rivers (Chenab, Jhelum, Indus)' (Wirsing and Jasparro 2006: 3). But with the building of the Baglihar dam in Doda district of Jammu and Kashmir, the dispute gained a new dimension. While Pakistan's objection to the dam had the usual elements regarding water sharing, there was also a security dimension, which was arguably the more important consideration. Pakistan argued that the dam could, potentially, be used to flood large parts of that country. And in 2005 for the first time Pakistan invoked the arbitration clause of the 1960 treaty.

The use, or abuse, of technology too could cause conflict-enhancing change in a river basin. In a simple case the overuse of water-intensive technology in the upper riparian regions will leave less water resources available for the lower riparian regions. But the problem can take on an entirely more threatening dimension if the technologies have an adverse effect on the water source itself. A striking example is that of the Zeravshan River Basin in Central Asia.

> The Zeravshan River Basin includes large parts of northern Tajikistan and southern Uzbekistan, including the cities of Samarkand and Bukhara. Several decades ago it became a closed basin. The Zeravshan River no longer empties into the Amudarya River; for three decades the river has instead disappeared into the sands between Bukhara and the Amudarya.
>
> The headwaters of the Zeravshan River run through some of the major mining areas of Central Asia, where the Soviet mining industry practiced a range of unsound mining practices. As a result, mining wastes from exposed tailings, landslides, and polluting processing facilities continue to pollute the Zeravshan, especially with mercury, arsenic, lead, antimony, and molybdenum. While Uzbekistan contributes to this pollution, the river's primary conflict concerns Uzbekistan complaints about Tajikistani pollution (Sievers 2002: 371).

A variety of changes along the course of a river thus have the potential to generate conflict. But even as we recognize the wide range of the potential for disputes it does not imply that all change will necessarily lead to conflict.

Change as Opportunity

Given the fact that conflict has a way of attracting attention, it is easy to ignore examples of change that ease the degree of conflict, whether it is a technology that improves the efficiency of water usage or an overall process of development that reduces the number of persons dependent on water-intensive agriculture for a livelihood. The tendency to ignore the conflict-easing dimension of change can even make us quite sceptical about the opportunities that change can provide to reduce the pressure on the river. As a result some of the more striking successes in using the opportunities of change to overcome deep divisions are not always given their due. One such striking example of using change as an opportunity would be the case of the Mekong River. Beginning in the heights of the Tibetan Plateau, the 4,800 km long river flows through six distinct geographical regions. The Mekong River Basin includes parts of China, Myanmar and Vietnam, nearly a third of Thailand and most of Cambodia and Lao People's Democratic Republic (PDR). Beyond the sheer magnitude of this basin, it has attracted attention for less favourable reasons as well. The basin includes areas of widespread poverty; it includes the Golden Triangle which was, at one time at least, one of the major centres of global drug trade; and the wars in Vietnam, Cambodia and Laos made it one of the regions with the greatest man-made destruction.

The complexity and the pressures on the river basin attracted international attention as early as the mid-fifties, leading to the setting up of the Mekong Committee. The war conditions in the region however crippled this committee, severely limiting what it could achieve. But once peace returned to the region, there were a variety of steps taken to explore the possibility of transboundary cooperation along the river basin. The early steps were tentative and focused primarily on using the river as a means of improving connectivity. The river itself was not the easiest to navigate. But by the early nineties, the Chinese, relying heavily on the skill of their crews, managed to create a network of boats that regularly navigated the river. This opened up fresh potential for trade across the boundaries the river crossed. This was particularly useful for some of the poorer nations to reach across their boundaries, and sometimes find access to the much richer markets elsewhere on the river.

By the mid-nineties Cambodia, Lao PDR, Vietnam and Thailand saw the potential for jointly managing their shared waters and for the sustainable development of the Mekong River Basin. They recognized the need for institutions that would help develop this potential and set up the Mekong River Commission in 1995. But this institution had its inadequacies built-in. With the Chinese having developed, and continuing to dominate, the navigation of the river, a meaningful institution had to extend to at least those parts of China that had a direct stake in the river. More significantly, with globalization now creating new, effective global linkages, it was not enough to confine attention to the basin alone. There was now a potential to link some of the poorest regions in the world to ports that would, in turn, link them to the global market. The focus therefore had to move beyond the basin to the sub-region. This larger requirement is captured by the Greater Mekong Subregion. This sub-region consists of the Yunan province and the Guangxi Zhuang Autonomous region in the People's Republic of China, Vietnam, Lao People's Democratic Republic, Thailand, Cambodia and Myanmar. The focus on the sub-region as a whole meant that the cooperation could be stepped up to a level where the sharing of the river and its waters becomes just one part of a larger agenda. Indeed, the focus is on people along the river rather than river waters themselves. The river per se itself then becomes only a part of the four strategic pillars of the Regional Cooperation and Strategy Program of the Greater Mekong Subregion. The four pillars are '(i) strengthening connectivity and facilitating cross-border movement and tourism; (ii) integrating national markets to promote economic efficiency and private sector development; (iii) addressing health and other social, economic and capacity building issues associated with sub-regional linkages; and (iv) managing the environment and shared natural resources to help ensure sustainable development and conservation of natural resources' (ADB 2006). While the navigation and tourism aspect of the Mekong River form a part of the first pillar, the river itself and its water resources are a part of the fourth pillar. The structural part of the program as captured by the second and third pillars focus more on meeting the needs of the people along the river.

There are several aspects of the experience of the Greater Mekong Subregion that could be useful when addressing the challenges that have emerged around other rivers. It emphasises the need to focus on the distinction between river resources that are potentially conflict generating and those resources that can only be tapped through cooperation. No matter how equitable and effective a water-sharing arrangement may be, it would always be sensitive to changes in water use on either side. There is thus always potential for conflict in areas like sharing irrigation waters. On the other hand, there are resources whose potential only exists when they are tapped by more than one interest group and whose value can in fact increase the more it is tapped. The Greater Mekong Subregion realized quite early that navigation was one such resource. It only had value if the connectivity it provided was used by more than one country. And the more it was used the greater its value. Over the years the countries along the Mekong have begun to recognize that tourism too has a similar effect. By building on areas of cooperation they are today in a stronger position to address disputes over water sharing as and when they arise.

The second critical aspect of the Greater Mekong Subregion is that it has been quick to recognize that the full potential of some of the economic impulses it has tapped into can only be realized if it is not confined to the river basin. The interests of people along the river can extend beyond the river basin. In geographical terms this need to extend the area of analysis is reflected in the shift from the Mekong River Basin to the Greater Mekong Subregion. But the need to extend original impulses can be seen in other aspects of the sub-region as well. The original connectivity provided through navigation along the Mekong River has been extended to major road corridors. A major part of the plans of the Greater Mekong Subregion consists of building north–south and east–west corridors that have already resulted in some of the dusty roads of the Vietnam War chronicles being turned into major expressways.

Institutions

With change along river providing both the potential for conflict as well as the opportunities for cooperation, the obvious challenge is to use these opportunities to overcome the pressures generated by conflict. The ability to tap opportunities in a way that can overcome conflicts would depend on the institutions that are given this task.

The term 'institution' is used here in a sense similar to that used by the philosopher John Rawls, who sees institutions as 'a public system of rules which defines offices and positions with their rights and duties, powers and immunities, and the like. . . . As examples of institutions, or more generally social practices, we may think of games and rituals, trials and parliaments, markets and systems of property' (Rawls 1971, 1999: 47–48). Such a definition is broad enough to include all official and unofficial mechanisms to deal with the river. Within the official mechanisms it would include the entire range from river water treaties to river basin authorities. And among the unofficial mechanisms would be traditional arrangements for water sharing.

In analyzing, let alone evaluating, river water institutions we invariably come up against a common problem: an institution may be created with one set of objectives and the rules that are consistent with those objectives; but in reality it may function as something else. A river basin authority can be created to implement a particular formula for water sharing but it may find itself unable to do so and, in extreme cases, it may even have to witness principles of water sharing that are quite contrary to what it was set up to implement. In addressing this dilemma too we would do well to be guided by Rawls:

> An institution may be thought of in two ways: first as an abstract object, that is, as a possible form of conduct expressed by a system of rules; and second, as the realization in the thought and conduct of certain persons at a certain time and place of the actions specified by these rules. There is an ambiguity, then, as to which is just or unjust, the institution as realized or the institution as an abstract object. It seems best to say that it is the institution as realized and effectively and impartially administered which is just or unjust. The institution as an abstract object is just or unjust in the sense that any realization of it would be just or unjust (Rawls 1971, 1999: 48).

This view of an institution being what it does rather than what is stated in any given concept is particularly useful when dealing with river basins. What goes under the name of river basin organizations can in fact cover a very wide range of concepts of institutions. The role provided for different stakeholders varies from one organization to another. Some have a prominent role for the local farming community, while others may barely have any representation for them. By the same token an institution that would be able to tap the opportunities provided by change to overcome the conflicts change throws up would have to be judged by how it performs this task. The precise features of such an institution would vary substantially as it has to be sensitive to the requirements of each river basin. But there would seem to be at least one fundamental characteristic that such an institution would need in order to be effective.

The institution would require a large enough canvas to tap the opportunities available as well as to enable river-based initiatives to realize their full potential. As the Greater Mekong Subregion has shown there is reason to look beyond the river basin to the sub-region and even further. But the expansion of the institution to cover a larger geographical region than the basin can bring in people whose interest in the river basin itself may be secondary. Some of these groups may even prefer to, say, keep a conflict over water sharing going as a source of pressure, since they themselves have nothing to lose from such a position. In order to avoid such a situation we need to focus on the people of the basin rather than the basin as a geographical entity. The boundaries that need to be drawn would then cover the interests of the people in the basin, wherever those interests may lie, whether in the basin or outside. At the same time the focus on the primary concerns of the people in the basin will also ensure that persons whose primary concern is not within the basin itself will not have a dominant role to play. In other words, while the canvas will include all the opportunities, both within and outside the basin, that are available to be tapped for the benefit of the people of the basin, issues that are not considered of primary importance to them can be kept out. Such a boundary will be defined not so much in geographical terms alone, but will also have different economic, social and cultural limits. The economic boundaries would cover all the specific activities anywhere in the world that can help or harm the basin. The social and cultural dimension too would extend to any part of the world that the river basin can relate to. Ideally, the political dimension would allow for the basin's interests to be protected and developed vis-á-vis other groups in the polity.

Conclusion

Our quick overview of disputes over river waters across the world points to the value of taking on board a view that goes beyond the details. Such a view makes it evident that the problem is not one of water scarcity alone, or even changes in the river basin alone, but one of the institutions not being able to adapt to changes along the river. It is then futile to assume, as is often done implicitly, that the changes along the river are minor or can be reversed; or that existing institutions are always best suited to deal with the changing ground realities. This is particularly true if we recognize that technological change and the forces of globalization have increased both the scope for change as well as the challenges before institutions that have to deal with them. The way forward must then include an evaluation of changes that have taken place along the river and the institutions that have to deal with them. And the role and functioning of the institutions would have to be defined in a way that focuses on the people of the river basin, rather than the river waters alone. It is not entirely clear that such an exercise has been carried out in the case of the Cauvery.

References

ADB. 2006. 'The Greater Mekong Subregion — Beyond Borders (2007–2009) Regional Cooperation Strategy and Program Update'. Asian Development Bank. Available at http://www.adb.org/Documents/CSPs/GMS/2006/csp0100.asp

CWDT. 2007. *The Report of the Cauvery Water Disputes Tribunal with the Decision* — Vol I. New Delhi.

HDR. 2006. *Beyond Scarcity: Power, Poverty and the Global Water Crisis.* Human Development Report. New York: United Nations Development Programme.

McCool, Daniel. 1993. 'Intergovernmental Conflict and Indian Water Rights: An Assessment of Negotiated Settlements', *Publius*, 23 (1): 85–101.

Nair, Janaki. 2005. *The Promise of the Metropolis: Bangalore's Twentieth Century.* New Delhi: Oxford University Press.

Pani, Narendar, Tara Srinivas and Vinod Vyasulu. 1985. 'Impact of Colonialism on the Economic Structure of Indian Cities: Bangalore 1800–1900', in Vinod Vyasulu and Amulya Kumar N. Reddy (eds) *Essays on Bangalore Vol 1.* Bangalore: Karnataka State Council for Science and Technology.

Rawls, John. 1971, 1999. *A Theory of Justice*. Oxford: Oxford University Press.
Sievers, Eric W. 2002. 'Water, Conflict and Regional Security in Central Asia', *NYU Environmental Law Journal*, 10.
Wirsing, Robert G. and Christopher Jasparro. 2006. 'Spotlight on Indus River Diplomacy: India, Pakistan, and the Baglihar Dam Dispute', Asia-Pacific Centre for Security Studies.
Wolf, Aaron T. 2000. 'Indigenous Approaches to Water Conflict Negotiations and Implications for International Waters', *International Negotiation*, 5 (2): 357–73.
———. 2004. *Regional Water Sharing as Confidence Building: Water Management as a Strategy for Peace*. Berlin: Adelphi Research.
Wolf, Aaron T., Shira B. Yoffe and Mark Giordano. 2003. 'International Waters: Identifying Basins at Risk', *Water Policy*, 5: 29–60.
Wolf, Aaron T., Annika Kramer, Alexander Carius and Geoffrey D. Dabelko. 2005. 'Managing Water Conflict and Cooperation', in *State of the World 2005: Redefining Global Security*. Washington: World Watch Institute.

4

Resolving River Water Disputes in India: Reflections

Ramaswamy R. Iyer

Introduction

At the outset, three preliminary points need to be made. First, this article will confine itself to water-sharing disputes, and not cover disputes involving environmental or pollution or security issues. Secondly, the widely held view that the existing system in India for the resolution of inter-State river-water disputes is unsatisfactory and needs an overhaul is somewhat exaggerated, though (as we shall see later) some changes might be warranted. The conflict-resolution machinery is creaking badly, not because it is badly designed but because we have wrecked it. The 'we' here means the State and Central Governments, politicians, lawyers, water-users, the media, the intelligentsia and the general public. Thirdly, we are excessively preoccupied with the details of particular disputes, and fail to see river-water disputes as a subset of the larger set of water-related disputes in general, and to ask ourselves what the root causes of such conflicts are. We shall return to the second and third points, but first let us take note of the existing legal framework in broad outline.

Legal Framework

The principal components of the legal framework are (*a*) Entry 17 in the State List which is the primary entry relating to water in the Constitution; (*b*) Entry 56 in the Union List which gives a potential role to the Central Government in relation to inter-State rivers to the extent that Parliament legislates for the purpose (and the River Boards Act 1956 enacted by Parliament under that Entry); (*c*) Article 262 which provides for the adjudication of inter-State river-water disputes; (*d*) the Inter-State Water Disputes Act 1956 enacted by Parliament under that Article; and (e) the amendments made to the ISWD Act in 2002.

Let us forthwith get one red herring out of the way. There has for long been a view that the configuration of entries relating to water in the Constitution is inappropriate and needs to be changed; that under this configuration the Centre is ineffective in relation to water; and that in order to enable it to play a proper role, water should be put into the Concurrent List. When the Constitution was being drafted, there was perhaps an arguable case for putting water into the Concurrent List, but such a shift at this stage seems very improbable, as it would go counter to the whole trend of the last decade or more towards decentralization and fuller federalism. It is also quite unnecessary. What an entry in the Concurrent List implies is that both the Centre and the States can legislate on the subject. However, the Centre can legislate now on water under Entry 56, but has made little use of that enabling provision; and the River Boards Act 1956, which it did enact, remains a dead letter. Instead of pursuing the chimera of a constitutional amendment to shift water to the Concurrent List, the Centre could usefully explore the possibilities of legislation under Entry 56 and of reactivating the moribund River Boards Act. Incidentally, if the River Boards Act had been acted upon and River Boards set up for important inter-State rivers such as Krishna, Godavari or Cauvery, dispute-settlement could have taken the alternative route of arbitration under that Act. Without River Boards, that remains only a hypothetical possibility. However, it is not entirely clear that arbitration under the River Boards Act would have been significantly different from adjudication under the Inter-State Water Disputes Act.

Policy Framework

From the legal framework, let us turn to the policy framework. The National Water Policy 1987 was the outcome of the first attempt to bring about agreement among the States on a minimal set of basic statements about water. That exercise was difficult enough without being complicated further by trying to incorporate a statement on the highly contentious subject of inter-State river-water disputes. The NWP 1987 therefore steered clear of the subject. The new NWP 2002 also confines itself to bland generalities on inter-State water-sharing. Separately, the Ministry of Water Resources did attempt a draft statement of water-sharing principles and the draft went up to the National Water Resources Council once or twice in the 1990s, but the wide divergence of views among the States made it a non-starter; it remains in limbo. There seems to be little likelihood of an agreed statement emerging in the foreseeable future. Meanwhile disputes have to be dealt with, and successive Tribunals have referred to the Helsinki Rules, case-law in India and elsewhere, reports of earlier Tribunals, and so on. By and large, the principle adopted by the Tribunals is that of equitable apportionment. Even if a national statement on water-sharing had been agreed upon, it could hardly have laid down any principle other than that of equitable apportionment or sharing, and it would necessarily have been a very general statement needing to be elaborated in detail in each case. A National Inter-State Water-Sharing Policy Statement would doubtless have been useful, but its absence is not a serious constraint; there are enough principles to guide us.

Adjudication Processes: Criticisms

Broadly speaking, there are four main criticisms of the prevailing adjudication process for the resolution of inter-State river-water disputes.

1. Adjudication is not the appropriate means of settling such disputes. A negotiated agreement, with the assistance of mediation or conciliation if necessary, is the best way.
2. The adjudication system under the ISWD Act is dilatory and cumbersome. There are delays at every stage.
3. The proceedings are adversarial and divisive. Each side engages eminent counsel for arguing its case strongly, makes maximal claims, and fights every inch of the way. The procedure precludes a problem-solving approach, or any effort towards the composition of differences. Under this system, the parties have to play the role of disputants, and the responsibility for resolution is solely that of the judges.

4. When finally a decision is given, there are no effective means of ensuring compliance with it. Moreover, one or more parties may be left with a sense of grievance or injustice, for which there is no remedy.

The following is a brief discussion of these points.

1. It is indeed true that agreement is better than adjudication. However, Article 262 and the ISWD Act do not force adjudication on the disputing parties, nor do they preclude recourse to negotiation, conciliation or mediation; but when all these efforts fail, disputes still have to be resolved, and a last-resort mechanism is needed for the purpose. That is what Article 262 and the ISWD Act provide. We are not compelled to invoke them, but they are available if all else fails. Such a last-resort mechanism is very necessary, and we should be grateful that we have it. Let us remember that in the Cauvery case two decades of negotiations failed to produce results, and it was only thereafter, and under the directions of the Supreme Court, that the Tribunal was set up. After that, we have to try and make adjudication work rather than keep repeating that agreement is better than adjudication.

2. Delays at every stage certainly presented a serious problem in the past. The Sarkaria Commission made some recommendations in this regard, and after prolonged consideration, they have been implemented through the Amendments of 2002. Now the Central Government has to establish a Tribunal within a year after a State Government asks for one. (It may be noted that this still leaves a year for further explorations of a negotiated settlement.) The Tribunal has to deliver its award in three years, but can seek an extension of two years, making a total of five years in all. Delays can still take place at the stage of a supplementary or clarificatory report by the Tribunal in response to a reference made to it within three months after the Award, or in the notification of the Order by the Government of India in the Gazette.

However, it seems probable that after the amendments of 2002, the problem of delays at various stages is likely to be substantially diminished.
3. Adversarial proceedings characterize all litigation in the courts, and as ISWD Tribunals function like courts, their proceedings too are subject to this malaise. However, there is no law against a constructive, cooperative approach to adjudication. The proceedings can be substantially different. We shall return to this.
4. The problem of non-compliance is indeed a serious one. Though the Award of an ISWD tribunal is said to be final and binding, there are no means of enforcing compliance with it. If a State Government refuses to obey the Order of such a Tribunal, there are not many courses open to the other parties to the dispute or even to the Central Government. The Centre can give directions, but if these too are not complied with, what sanctions are available? Article 356 (Central rule) is an extreme measure that cannot be lightly used, and in any case, what will happen when a popular government returns? The Sarkaria Commission had recommended that the words 'final and binding' in the Act should be buttressed by conferring upon the Tribunal's Order the status of an Order or Decree of the Supreme Court, and this has been done through the 2002 amendment. However, this seems to have had no perceptible effect.

Some Substantive Issues

(a) Taking groundwater into account

Apart from the procedural and operational aspects discussed above, ISWD Tribunals have also been criticized for confining themselves to the river in question and ignoring groundwater. This is an important enough matter to warrant consideration. Undoubtedly, surface water and groundwater are inter-linked and one cannot be considered in isolation from the other. However, disputes whether between States in India or between countries (in this sub-continent or elsewhere) do not arise over water-resource endowments as a whole. States and countries accept differential water endowment as a fact of nature. However, they do quarrel over rivers. This could be because rivers are very visible, and they flow from State to State and country to country (and the impacts of interventions on rivers are very visible and real too); rivers are felt to be part of the lives of the people; people have a strong relationship with rivers (for instance, both Karnataka and Tamil Nadu have a strong sense of identification with the Cauvery, and the river figures prominently in both Kannada and Tamil literatures; similarly, all the countries of the Nile basin have a strong sense of identification with the Nile); ways of living develop in relation to the river and the use of its waters (for instance, irrigated agriculture and paddy cultivation over the centuries in the Cauvery Delta formed the basis of a distinctive way of living); and so on.

The point is that disputes do develop over river waters — Cauvery, Krishna, Godavari, Ganga, Indus, Mekong, Nile, Rio Grande, and so on — and some kind of understandings over sharing become necessary on rivers and not on water resources in general. (That is why international documents emerge on river-water-sharing: the Helsinki Rules, the UN Convention of 1997). Theoretically, in a dispute between countries A and B, it may be logical for country A to say to country B 'You have other water sources, leave this river to us and refrain from asking for a share in it', but in practice such an approach will not work; a sharing of waters in the river in question will be necessary. (Similarly, theoretically, it may be possible for country B to say to country A 'We are better placed for agriculture than you are, so let the river flow to us, and we will produce food for you too; for your part, you might try and go in for other kinds of development, such as industry'. Such an approach also will not work: both countries will want to develop both agriculture and industry.)

In the negotiations over the Indus (India–Pakistan) and the Ganges (India–Bangladesh), groundwater endowments did not figure. Coming to the internal context, the ISWD Act is about disputes over river waters, and the Tribunals set up under it are mandated to deal with the river in question and not with the respective overall water resource endowments of the States concerned. (The overall endowments are dealt with by bodies such as the National Commission on Integrated Water Resources Development Plan (Report, 1999), but they merely make estimates and not mandatory allocations of total water resources.)

Having said that, it must be acknowledged that in making allocations of river waters to the disputing States an ISWD Tribunal must take into account the availability of water from alternative sources, including groundwater, as one of the relevant factors, though not the sole or determining one. Among the many criteria laid down in the Helsinki Rules for equitable apportionment for beneficial uses, the availability of water from alternative sources is one, as also the extent to which the area in question is dependent on the river that is in dispute.

(b) Fragmentation

Even if the principle of equitable apportionment for beneficial uses is properly and fairly applied, conflict-resolution through allocations of the waters to the various disputing parties still suffers from the defect of fragmentation. Each State is given a 'share' in the river waters and left free to use it as it deems fit. This involves a segmentation or fragmentation of the river system, which is perhaps not the ideal course. It would doubtless be better for all the riparian States to come together and jointly use the system in an integrated and holistic manner. However, this requires a degree of reasonableness and wisdom that is not always available. If the best course is not feasible, then we have to settle for the second best. Disputes do need to be settled, and accord is better than discord. From this point of view, allocations of river waters, whether by agreement or by adjudication, seem a sensible means of resolving inter-State river-water disputes.

(c) Projects as foci of conflicts

Incidentally, it is interesting to note that conflicts over river waters, whether inter-country or intra-country, rarely arise so long as the rivers are flowing naturally, unimpeded by human intervention; they seem often to arise in the context of large projects. It would appear that such projects tend to become foci of conflicts. This is essentially because they tend to alter geography and hydrological regimes, sometimes drastically; and they involve issues of control, power and political relations, social justice and equity. However, that complex subject cannot be discussed in this article.

Suggested Legal Reforms

a) Bringing Disputes within Supreme Court's Original Jurisdiction

A legal reform that has been suggested by the National Commission to Review the Working of the Constitution (and also by the eminent lawyer Fali Nariman) is that the ISWD Act should be repealed and inter-State river-water disputes brought within the original jurisdiction of the Supreme Court. This arises largely from a sense of exasperation with the manner in which adjudication under the ISWD Act has been functioning. However, will the Supreme Court be able to cope with the enormous burden that this will caston it? At the moment, there is the possibility of several such disputes being heard in parallel by different tribunals. If all of them have to be dealt with by the Supreme Court (in the exercise of original jurisdiction), will it take them up sequentially, or will it constitute several Benches to deal with them simultaneously? How feasible is the latter course? Besides, there is a danger that this will become the principal work of the Supreme Court, crowding out large numbers of other categories of important cases. Nor is it certain that proceedings in the Supreme Court will be speedier than in the Tribunals. Time-limits have been set for the Tribunals by the 2002 amendments; who will impose time-limits on the Supreme Court? One cannot be confident that a dispute which might take 5 years before a Tribunal will be disposed of in 2 years by the Supreme Court. Even on the question of better compliance with the final decision, there are grounds for a degree of scepticism. It may be recalled that a former Chief Minister of Karnataka was at one stage prepared to defy the Supreme Court, and take the chance of contempt proceedings; and that Punjab passed an Act terminating all past water accords, seeking to destroy in the process the very basis of certain directions of the Supreme Court. It is of course true that in the former case good sense prevailed at the last stage and the Chief Minister pulled back from the brink; and that in the latter case, the Punjab action has been referred by the Central Government to the Supreme Court for an opinion, which is still awaited. However, the point is that implicit obedience to the Supreme Court's Order cannot be taken for granted, though it is perhaps less likely to be defied than an Order of a Tribunal.

The suggestion of bringing inter-State river-water disputes within the ambit of the Supreme Court's original jurisdiction is based on the position that prevails in the USA. However, when such a dispute goes before the Supreme Court of the United States, what the Supreme Court does is to appoint a Master forthwith, and the Master proceeds to go into dispute in great deal, holds hearings and arrives at findings and recommendations, which are then placed before the Supreme Court. The role that the Master plays is analogous to that of the Tribunal in our system, though the style is different. Some such machinery is clearly necessary.

The repeal of the ISWD Act and the transfer of inter-State river-water disputes to the original jurisdiction of the Supreme Court is perhaps a counsel of despair that is not called for. Our system seems better suited than the American one to our conditions and circumstances, and it can surely be made to work better.

b) Providing for an Appeal to the Supreme Court

The most important deficiency in our system is that the bar on the jurisdiction of the Courts, provided for in Article 262 and incorporated in the ISWD Act, makes the Tribunal's decision a single, non-appealable verdict. If one or more parties are left with a sense of grievance or injustice, as can happen and has in fact happened in the Cauvery case, the aggrieved party has no remedy apart from a reference back to the same Tribunal within three months. (I know that in that case petitions have been submitted to the Supreme Court; we shall discuss that development later.)

A recommendation that I have been putting forward for several years though it has received hardly any attention, is that in partial modification of the bar of jurisdiction of the courts, the ISWD Act should be amended to provide for an appeal to the Supreme Court against an ISWD Tribunal's Order. One knows all the arguments against this: every case will go to the Supreme Court; the cases will get further delayed; and so on. The answer is that every case does go to the Supreme Court even now, on some issue or the other, and the Supreme Court rarely ever says: 'We do not have jurisdiction; go back to the Tribunal', nor does it carefully refrain from entering into water-sharing issues; it is clear that the parties do wish to go the highest Court in the land, and it seems better to accommodate that wish. As for delays, the 2002 Amendments will cut down the time taken at different stages, and we can afford to provide a year or two for an appeal to the Supreme Court. It will be eminently worthwhile, if it assuages to some extent the sense of grievance felt by one or more parties. Further, as mentioned earlier, this might improve the prospects of compliance to some extent, though this carries the sad implication that compliance with the Tribunal's Order cannot be presumed, whatever the ISWD Act might say. It is submitted that my suggestion combines the advantages of our system of Tribunals with those of the US system of final decision by the Supreme Court.

In parenthesis, a further possibility may be mentioned: that of replacing the present system of setting up a Tribunal in each case by a Standing or Permanent Inter-State River-Water Disputes Tribunal with multiple benches, if necessary. If such a course were adopted, other features of the present system, namely, a request by a State Government to the Central Government, a year's time for exploring negotiations further, the reference of the dispute to the Tribunal by the Central Government, etc., should remain; and as argued above, provision should be made for an appeal to the Supreme Court against the decision of the Tribunal.

Change of Style

The other major problem of the present adjudication system is the adversarial nature of the proceedings, based as they are on court proceedings. However, Tribunals are not obliged to adopt the court style of functioning. Instead, they can come down from their lofty position above the disputants and sit at a round or oblong table with them, and adopt a consultative, interactive, fact-finding, solution-exploring Committee-style procedure, while retaining the ultimate responsibility for giving a judicial decision. This has been suggested by Fali Nariman, and there is much merit in it.

Bringing the People in

Such a consultative, interactive approach to conflict-resolution must also be an inclusive one: it must include and involve those who have a vital interest in the decision, namely the farmers, industrial establishments, municipalities, and people in general because all of us are water-users. However, such inclusiveness is not a feature of the present system. In terms of the ISWD Act, 'inter-State' really means 'inter-government'. The disputing parties under the Act are the State Governments concerned and not the people. The Tribunal does not hear the farmers and other water-users in the basin. The MIDS initiative of bringing the farmers of the Cauvery Basin in Tamil Nadu and Karnataka together, and the emerging concept of 'Cauvery Family', are now well-known. However, the Cauvery Family has no direct access to the Tribunal. It seems very desirable that any reform of the present system of resolution of inter-State river-water disputes should bring the people in as interested parties. 'People' in this context should encompass different categories of water-users, as also those who are likely to be affected by the projects that the Tribunal takes note of. In the Narmada case, the Narmada Water Dispute Tribunal not merely allocated the waters of the river but also mandated certain crucial technical features of two projects, one in Gujarat and the other in Madhya Pradesh, and even laid down the norms and conditions for the rehabilitation of people likely to be affected by the projects; but those people themselves were not parties before the Tribunal. This needs to change.

Making the Existing System Work

Pending those reforms, we have to make the existing system work as efficiently and effectively as possible, and we can do so, provided we respect the system and not wreck it as we have been doing. Consider the following: adjudication is deprecated and resisted as far as possible; an Order of the Tribunal is sought to be nullified by an ordinance, until the latter is struck down by the Supreme Court; aspersions are cast on the Tribunal; a reconstitution of the

Tribunal is called for at a time when its work is in the final stages; the Supreme Court is defied; when, after many years of hearings, listening to elaborate arguments by learned counsel, and studying the massive documentation submitted by all parties, the Tribunal pronounces its Final Order, it is immediately denounced as unjust and discriminatory; resolutions are passed rejecting the Order, though in law it is final and binding and has the force of an Order of the Supreme Court; the lawyers go on strike; an organization, represented by a distinguished legal luminary, submits a petition to the Supreme Court on certain grounds, and in an act of grave irresponsibility, prays not for relief or remedy but for the quashing of the Tribunal's Order, seeking to nullify seventeen years' proceedings; a former Prime Minister of India lambasts the Tribunal's Order in strong language and at the same time makes a sanctimonious plea for national integration; a grievance against the Tribunal leads to negative feelings about a people and their language; and so on.

These negative activities might not actually have done much harm, but the underlying attitudes are deeply disquieting. If we are ready to flout the law of the land, declare that we and not the judges will decide what is just or unjust, and have no respect for constitutional mechanisms, no regard for federalism, and no concern for good relations with neighbouring States, then neither the Tribunal system nor any alternative arrangement for conflict-resolution will work. This is what was meant by the reference at the beginning of this paper to 'wrecking the machinery'. Having done our best to wreck the machinery, we then criticise it for ineffectiveness, ignoring the fact that we have contributed to that ineffectiveness.

One is not singling out Karnataka for criticism. There is hardly any doubt that if the allocations in the Cauvery Tribunal's Final Order had been different and had caused serious disappointment in Tamil Nadu, reactions and responses in that State might not have been very different from what we have seen in Karnataka.

Let us consider a contrast. The decisions of the Neutral Expert appointed under the Indus Treaty on the differences between India and Pakistan over the Baglihar Project were not wholly in favour of either country. Pakistan's serious objections to the project design were not upheld; on the other hand, India has been asked to make changes in a few features. Neither side has denounced the decisions as unfair or biased. Pakistan has not reacted angrily to the NE's Report; and India has not refused to carry out the recommended changes. The responses of both sides have been muted, responsible and civilized. It is interesting to compare this with the responses of the State Governments to the Cauvery Tribunal's Report.

One is not saying that dissatisfaction with the Tribunal's Final Order in any or all of the States is illegitimate. There are means available for the expression of that dissatisfaction, namely a petition to the Tribunal; and all Parties have filed their petitions. That is perfectly in order; what is not in order is the kind of comment and response that was referred to earlier.

Bar of Jurisdiction of the Courts

Let us turn now to the Special Leave Petitions (SLPs) that Karnataka, Kerala and Tamil Nadu have filed before the Supreme Court. It is somewhat surprising that the Governments, their eminent Counsel and even the Supreme Court have chosen to ignore the explicit and specific bar of the jurisdiction of the courts provided for in Article 262 and enacted in the ISWD Act. One would have expected the bar to be taken note of, and the question whether it can be over-ridden and SLPs under Article 136 admitted to have been carefully discussed. The SLPs seem to have been forthwith admitted without discussion, as if the bar of jurisdiction in the ISWD Act as enabled by Article 262 did not exist. It is difficult to understand the implicit consensus on the part of the Governments, their eminent Counsel and the learned judges that express provisions of the Constitution and the laws can be ignored as non-existent or as of no consequence. Incidentally, assuming that a way can be found in the Cauvery case for bypassing the bar of jurisdiction and reaching the Supreme Court through certain provisions of the Constitution (Articles 32 or 131 or 136), one has to ask: under what circumstances will the bar actually operate? If a route to the Supreme Court through one or more of those Articles is available in this case, it will surely be available in all cases, and the bar will become virtually non-existent. Would that not amount to a repeal of parts of Article 262 and the ISWD Act? Speaking speculatively and with great respect, one wonders whether the Supreme Court is perhaps unhappy at the bar and pre-disposed to accept petitions.

There seems to be a degree of ambivalence not merely on the part of the Executive but even on that of the Supreme Court towards the ISWD Tribunals. It may be recalled that in the 1990s, when Tamil Nadu went to the Supreme Court asking for a pronouncement on the status of a Tribunal's Order, the Supreme Court chose not to answer the question and eventually obliquely persuaded Tamil Nadu to withdraw its petition. There seems to be a general tendency to denigrate the adjudication system provided by Article 262 and the ISWD Act, and speaking subject to correction, and again with deep respect, the Supreme Court has not been exempt from that tendency.

However, regardless of one's surprise at the unargued admission of the SLPs, one is in a way glad that the case is before the Supreme Court, because there will now be no room for a sense of grievance or injustice after the Supreme Court has heard the Petitions and given judgment. One must hope that the Supreme Court's verdict and the Tribunal's Supplementary or Clarificatory Order will not be greeted with emotional outbursts in any of the States.

Root Cause of the Conflicts

Finally, we may return to the point made at the beginning that inter-State river water disputes are only a sub-set of the larger set of disputes over water resources in general, and indeed over all natural resources including water. The root cause of such conflicts is a competitive unsustainable demand for water. The demands of the various parties to a dispute cumulatively add up to more water than exists. This leads to conflicts, as also to proposals for bringing in external water from distant sources. Supply creates demand and necessitates more supply. The availability of water leads to the adoption of water-intensive cropping patterns. More water is needed even to continue with this kind of agriculture; and of course, there is a desire to expand that agriculture under the imperative of 'economic growth', creating a demand for still more water, until the demand becomes unsustainable. There is always a demand for more water and still more water. So Karnataka and Tamil Nadu fight over the Cauvery, and Punjab terminates all water accords.

But where will this 'more water' come from? It has to be brought from somewhere. So big dams, canals and long-distant water transfers are planned. These will in turn generate new conflicts. It is clear, then, that what lies at the heart of water conflicts is 'greed' in Mahatma Gandhi's sense. Agreements, accords, treaties, and adjudications may temporarily bring peace, but the conflict will erupt again unless we learn to re-define 'development'. However, that is a much larger theme that will call for a separate article.

5
Regulatory Aspects in Water Resources Development and Management

R. Jeyaseelan

Introduction

Water is a precious and much-needed resource, but its availability is finite and highly variable in space and time. Its development and management is of prime concern for any country or region. The economic prosperity and food security of India has been linked with agricultural development which in turn depends on water resources. With the fast changing economic scenario (that is characterized by high rates of economic growth in India), change in world environment and lifestyles, development needs, industrialization, urbanization, and for other requirements such as ecology, competing demands for water has increased manifold. Available supplies are under great duress as a result of high population growth, unsustainable consumption patterns, poor management practices, inadequate investment in infrastructure, lack of maintenance of systems and low efficiency in water use. Water issues have assumed greater importance, and are becoming more and more sensitive and complex. According to the National Water Policy, water is a prime natural resource, a basic human need and is a national asset, and its planning, development and management need to be governed by a national perspective. According to it, the vision for the water sector in India should envisage: 'Optimal sustainable development, maintenance of quality and efficient use of country's water resources to match the growing demands on this precious natural resource with active involvement of all stakeholders in order to achieve accelerated, equitable economic development of the country'.

Water Availability

With only 2.45 per cent of land resources and about 4 per cent of fresh water resources, India has to support 16 per cent of the world's population and 15 per cent of its livestock. Out of about 4000 BCM of precipitation in a year, as much as 3000 BCM comes as rainfall in a short monsoon period of 3–4 months from June to September. Even this occurrence is not uniform, is highly uneven, and occurs within a space of a few days marked by heavy rainfall. Average annual water resource potential is estimated to be 1869 BCM. But due to hydrology, topography and geological limitations, only 690 BCM of surface water can be utilized by conventional storage and diversion structures. Total replenishable ground water potential is estimated to be at about 433 BCM, making the total utilizable water potential as 1123 BCM. Out of this about 40 per cent is confined in the Ganga–Brahmaputra–Meghna system. Present level of utilization is 605 BCM out of which irrigation accounts for over 500 BCM. The utilization for domestic, industrial, energy and other sectors is of the order of 30 BCM, 20 BCM, 20 BCM and 34 BCM, respectively. While assured utilizable water of the country is of the order of 1123 BCM (including ground water), water needs for the year 2050 has been assessed as about 1447 BCM (this can be reduced by technological innovations and improved efficiencies in all sectors of water use), indicating a substantial deficit. This calls for developing all possible sites to store the available utilizable surface water. Total storage created so far is 213 BCM from completed projects and about 76 BCM will be added by projects under construction, whereas the total required is about 460 BCM. In terms of per capita storage created, India has only 210 cum as compared to 1,110 cum created by China with a comparable population. In case corrective measures are not taken at this stage for appropriate regulation and improvement of efficiencies, the available water will not be able to meet the water demands. One of the factors leading to inefficient management of water resources had been the lack of involvement of stakeholders and regulatory measures.

In view of population growth, declining per capita water availability is a cause of serious concern. Though from overall country level scenario India may be above the internationally accepted standards of water scarcity, the per capita water availability at the basin level varies widely from 13,636 cum per year in the Brahmaputra–Barak basin to 298 cum per year in the Sabarmati basin. The situation is projected to get even more serious in 2050 when, about 22 per cent of the area and 17 per cent of the population in the country may be under absolute scarcity condition. The signal of the widening gap between water availability and water demand is evident from the projections made for the coming decades for different sectors such as food production, industries, etc. Catering to this ever-increasing, multi-sectoral water requirement equitably will be the most stringent challenge in the days ahead. While water is becoming scarce, the threats posed by floods and droughts coexist, getting more critical due to threats of climate change, as evidenced in recent times. It has become inevitable to search for new techniques, technologies and innovations for increasing the supply of 'usable water'. All these essentially aim at storage, diversion, conservation and recycling of utilizable water resources, and aim to prevent wasteful use and losses in systems.

Sustainability of Development

Sustainability demands utilization of the resources in the best possible manner without encroaching upon the right of the future generations to utilize the same resources for their sustenance and growth. Any skewed development pattern, which caters to a limited section of the society, cannot be termed as sustainable since it creates an undesirable unevenness. In case of the water sector, the issue of sustainability is even more pronounced since the human civilization as well as the entire flora and fauna rely heavily on the availability of water for survival and growth. The need of the hour is therefore an Integrated Water Resources Development and Management (IWRDM) programme backed by a strong multi-disciplinary institutional framework that primarily optimizes the allocation of the resources based upon some pre-determined, yet relatively flexible, priorities, and second, devises a suitable modus operandi for effectively utilizing the allocated resources.

Constitutional Provisions with Respect to Water

List I — Union List

Regulation and development of inter-state rivers and river valleys to the extent to which such regulation and development under the control of the Union are declared by Parliament to be expedient in the public interest by law.

List II — State List

Water that is to say, water supplies, irrigation and canals, drainage and embankments, water storage and waterpower subject to entry 56 of List I.

Any option of (i) transfer of water from State List to the Union List; or (ii) State List to the Concurrent List would require suitable constitutional amendment which appears to be a difficult proposition under the prevailing circumstances.

The NCIWRDP (1999) had gone into the constitutional and legal issues pertaining to the subject 'water' and had concluded that there is no need for any change in the provisions of the constitution, but of urgency was the need for the union government to pass laws more effectively under the existing constitutional provisions to deal with inter-state rivers. Accordingly, under Entry 56 of Union List, the Ministry of Water Resources could enact an act for inter-state rivers and river valleys (integrated and participatory management).

National Water Policy

In order to regulate and guide the development and management of water, Government of India adopted the National Water Policy in the year 1987, which was subsequently modified in 2002 and has been circulated to all the states. Priority of water use has been enumerated as a) drinking water, b) irrigation, c) hydropower, d) ecology, e) agro-industries, and non-agro industries, and f) navigation and other uses. Depending on the resource availability, development pattern and priorities, the state governments formulate their own state water policies taking into account all the emerging factors. The National Water Policy lays down principles for the wise and judicious use of water for survival of life, welfare of human beings and sustained as well as balanced growth. According to it, the basin is to be considered as the hydrological unit for development and management of water resources, starting from primary watersheds to sub-catchments, with the catchments integrated into sub-basins and basins.

Adjudication of Water Disputes

Article 262 of the Constitution of India dealing with the adjudication of disputes relating to water of inter-state rivers or river valleys states:

1. Parliament may by law provide for the adjudication of any dispute or complaint with respect to the use, distribution or control of the waters of, or in, any inter-state river or river valley.
2. Notwithstanding anything in this Constitution, Parliament may by law provide that neither the Supreme Court nor any other Court shall exercise jurisdiction in respect of any such dispute or complaint as is referred to in Clause (1).

Inter-state River Water Disputes Act (ISRWDA) 1956 was enacted under Article 262. Where negotiations do not lead to fruitful results, the water disputes are referred for adjudication to the Water Disputes Tribunal set up under the ISRWDA 1956. Based on the National Commission on Centre–State Relations' recommendations, necessary amendments in the existing Act of 1956 have been made and adopted as ISRWD (amendment) Act 2002. The amendments include a timeframe for the constitution of the ISRWD Tribunal, time limit for the tribunals to give their awards, notification and the decisions to have the same sanctity and force as an order or decree of the Supreme Court.

Experiences of the recent past in the resolution of water disputes through negotiation have been in some way futile due to the rigid stand taken by the states, and the process through tribunals have been protracted, slow and cumbersome, the mechanism considered by some as ambiguous and opaque. Even after adjudication by the tribunal, the cases get referred to the Supreme Court under Article 131 leading to further delays and non-implementation. Therefore, the dispute resolution mechanism needs rethinking and review.

Integrated Water Resources Development and Management

Sustainable development and management of water resources demands a holistic approach with due consideration of engineering, ecological, socio-economic and environmental aspects. Development of hydropower, management of floods and drought are to be integrated in the overall plan for equitable distribution, thereby minimizing the usual conflicts among multiple agencies working in the water sector, each with different goals, motivation and dynamics. To meet this objective, the strategy has to focus on augmentation of utilizable resources and conservation measures without sacrificing the standard of water quality. A suitable mechanism should cover systematic administrative and policy changes; bridging of the knowledge gap to enable informed decision-making, enactment of suitable legal instruments for enabling cooperation among various factions of society; appropriate economic and financial instruments ensuring sufficient flow of funds and educated involvement of all stakeholders.

Water Conservation

The present use of about 83 per cent of available water resources in the country by the irrigation sector can be progressively reduced to less than 70 per cent of all the water utilized to augment the growing demands for other uses such as drinking water, industries and ecological demands. Even a marginal improvement in efficiency in irrigation water use will result in saving large volumes of water which could be used for extension of irrigated area or for diversion to other beneficial purposes. In addition, to meet the increased food demands, it is imperative to store additional water. Water conservation could be achieved by creating storages, artificial recharge of ground water, reduction of conveyance losses, efficient water management by technology improvement, land treatments, conjunctive use of surface and ground water, re-use of waste water/return flows, improved operation and management of irrigation systems, rationalization of water rates, and the integrated use of poor and good quality water, etc. Conservation of rain water by suitable harvesting and recharge techniques/methods would augment water available for use. Industries need to resort extensively to zero-based budgeting, recycling and re-use as well as treatment of effluents to acceptable levels before discharging into water bodies for improving its re-use. Water budgeting and auditing are pragmatic options which can lead to water as well as cost savings.

Rainwater Harvesting and Artificial Recharge

The principle of rainwater harvesting is to conserve and retain water over a longer period of time by ground water recharge, and use the precipitation in the same area where it occurs. It minimizes erosion of precious top soil and results in better soil management. Rain harvesting methods comprise check dams, collection from rooftops, deep aquifer recharge, storage in soil profile, contour-bunding, small ponds around trees, percolation tanks, hydro-fracturing, etc., which are site specific depending upon the soil, topography, precipitation characteristics and climate. There are several success stories of usable water sources augmented substantially.

Recharging of ground water aquifer is of very great significance because it provides ready availability of water, minimal loss due to evaporation, protection against pollution, arrest of salt water intrusion in coastal belts and prevention of land subsidence in a depleted aquifer. In this regard, a number of ground water studies and conservation processes have been carried out by the Central Ground Water Board. The methods include spreading, recharge through injection wells, induced recharge from surface water bodies, and the conservation of sub-surface flows through construction of sub-surface dykes and percolation tanks.

Ground Water Legislation

The union government has circulated a model bill to the states and union territories to enable them to enact suitable legislation for regulation and control of ground water development, The bill was initially circulated in 1970, and was re-circulated in 1992 and 1996. While a number of states have enacted and implemented ground water legislation, some northeastern states feel that such legislation is not required, and some states are yet to respond.

Floods and Droughts

Out of the 329 Mha of geographical area, 40 Mha (45 Mha as per working group) is flood-prone, and 59 Mha is drought-prone. Floods have been amongst the major sets of disasters leading to frequent deaths and loss of property in numbers and amounts that have outnumbered any other natural disasters. The National Flood Commission (Rashtriya Barh Ayog) gave wide-ranging recommendation regarding the flood situation in its report during 1980. A review carried out by the Rangachari Committee during 2003 found that most of the recommendations had not been given any attention, and listed a few of them as priority areas, which has been circulated to the states for necessary follow-up and action. While it is an accepted fact that absolute flood control is not possible for any area/region, adequate level of protection could be provided by structural and non-structural measures. Flood-plain zoning is an important aspect in flood management, and a model bill was circulated to all the states by the union government in 1975, but the states are reluctant to enact the bill. None of the states have implemented the Flood Plain Zoning Regulation. For effective flood management, flood inundation map, flood hazard map, flood-risk zone map and flood-plain zoning map are very useful, but have not been taken up seriously. Flood insurance is also yet to take off, and has been slowed down by the absence of appropriate maps and regulations.

Drought affects the country year after year, and some areas have been identified as drought-prone. Measures need to be devised for drought forecasting, mitigation and new approaches for its management.

Environmental Aspects of Water Resources Development

A good environmental sense has been one of the features of India's ancient philosophy of development. Some of these concerns were integrated into various religious and social customs in such a manner that these were automatically taken care of by the people. Adequate provision for protection of environment and forest are made in the Constitution of India. Article 47 provides for protection and improvement of health. Article 48A is directed towards protection and improvement of environment and protection of forest and wildlife. Article 51(A) says it is the duty of every citizen to protect and improve the natural environment. Following the UN Conference on Human Environment (Stockholm, 1972) a constitutional amendment (42, 1976) inserted relevant provisions for environment protection in the constitution in Part IV — Directive Principles and Part IVA — Fundamental Duties.

In order to ensure sustainable development of water resources ensuring adequate protection of the ecosystem, the Government of India has enacted various acts and legislations. Prominent among these is the Environment Protection Act, 1986 through which the government has acquired wide powers for protecting the environment. Some other Acts related to water and environment are: Water (Prevention and Control of Pollution) Act, 1974 (amended in 1988); Water (Prevention and Control of Pollution) Cess Act, 1977 (amended in 1991); Forest Conservation Act, 1980; Environment Impact Assessment Notification of 1994 (amended in 1997) and the Ministry of Environment and Forest's (MOEF's) Notification of January, 1997 constituting the Central Ground Water Authority (CGWA), and MOEF's Notification of June, 2001, constituting the Water Quality Assessment Authority (WQAA).

The Water (Prevention and Control of Pollution) Act, 1974 seeks to maintain or restore 'wholesomeness of water' and the Central and State Pollution Control Boards have been set up under this Act. According to the Water Cess Act, 1977, both the central and state governments have to provide funds to the boards for implementing this Act. The Forest (Conservation) Act, 1980 provides for compensatory afforestation to make up for the diversion of forest land to non-forest use. The Environment (Protection) Act was enacted in 1986 for the protection and improvement of human environment. The EIA Notification of 1994 has made environmental clearances mandatory for all new projects, and for the expansion/modernization of existing projects covering 29 disciplines (later increased to 30) that include hydropower, major irrigation, and flood control projects. Its amendment in 1997 has made it mandatory to hold environmental public hearing before according environmental clearance.

The mandate of the Central Ground Water Authority is 'regulation and control of ground water management and development'. The authority promotes the conservation and protection of ground water, and the promotion of mass awareness. Its mandate is to adopt a proactive approach to sensitize water users and the general population with regard to the need for judicious use and scientific management of water which includes mass awareness programmes, rainwater harvesting, notification of critical areas, regulation and control of ground water development in notified areas, registration of persons/agencies engaged in construction of water wells, and clearance to ground water based industries. The CGWA can invoke any penal provision under the Environmental (Protection) Act, 1986. The following powers are given to the CGWA to perform the following functions:

(i) to exercise powers under Section 5 of the Environment Protection Act, 1986 for issuing directions and taking such measures in respect of all matters referred to in Subsection (2) of Section (3) of the said Act,
(ii) to resort to the penal provisions contained in Sections 15 to 21 of the said Act,
(iii) to regulate and control, management and development of ground water in the country and to issue necessary regulatory directions for this purpose, and
(iv) to exercise powers under Section 4 of Environment (Protection) Act, 1986 for appointment of officers.

Constitution of the 'Water Quality Assessment Authority' recognizes the importance of water quality monitoring through an extensive network at the national and state level, constitution of state level 'Water Quality Review Committees' in the country, standardization and unification of the process of monitoring.

Establishment of River Basin Organizations

The hydrological basin is generally seen as a desirable unit for the development and management of water resources, starting from primary watersheds to sub-catchments and, catchments integrated into sub-basins and basins, taking into consideration both surface and ground water for sustainable multi-sectoral use incorporating aspects of both quantity and quality as well as environmental considerations. All individual developmental projects and proposals should be formulated and considered within the framework of such an overall plan keeping in view the existing agreements/awards for a basin or a sub-basin so that the best possible combination of options can be selected and sustainably implemented. For this purpose, River Basin Organizations (RBOs) needs to be established. There are some models available the world over for guidance, but each RBO should be guided by the regional requirement, geographical, geophysical and soil conditions, climatic parameters, constitutional obligations and political considerations. Though there has been a general consensus on the constitution of the RBOs, the matter of defining their form and role is still under active consideration.

The National Water Resources Council under the headship of the Honourable Prime Minster of India and the National Water Board chaired by the Secretary, Ministry of Water Resources are in place to deal with major issues relating to the water sector. The issue was deliberated by a working group constituted for the purpose of deriving a mechanism for working out in detail the modalities/model(s) of RBOs. It was decided that a comprehensive institutional framework to ensure fairness in the basin-wise decision for optimum development and process of accountability supported by a strong and comprehensive but flexible regulations/legislation should be the backbone of the RBOs, and all the relevant bodies involved in the water sector must galvanize their individual ideas to form a concrete strategy for attaining the wider objective of sustainable development. Drawing up of appropriate 'Guidelines for distribution/sharing of interstate river waters amongst the riparian states' was assigned to a Working Group with the active participation of the states. Various awards of River Water Disputes Tribunals also include recommendations for formation of RBOs such as Krishna Valley Authority, Cauvery River Authority, etc. Even though the River Boards Act was enacted in 1956, nothing tangible has been achieved so far due to the varying views and reluctance expressed by the states.

Participatory Approaches

The National Water Policy (2002) states,

> Management of Water Resources for diverse uses should incorporate a participatory approach by involving not only the various governmental agencies but various aspects of planning, design, development and management of the water resources schemes. Necessary legal and institutional changes should be made at various levels for the purpose, duly ensuring appropriate role for women. Water users' associations and local bodies such as municipalities and Gram-Panchayats should particularly be involved in the operation, maintenance and management of water infrastructures/facilities at appropriate levels progressively, with a view to eventually transfer the management of such facilities to the user groups/local bodies.

Recognizing the need to provide legal back-up to PIM in the country, Ministry of Water Resources Commissioned an external organization (NGO) to suggest amendments in the existing irrigation acts which could be recommended to the states for incorporation in their respective State Irrigation Acts. Many action research programmes on the formation of Water User's Associations (WUAs) were initiated. The MoWR has been organizing national-level training programmes on PIM in various parts of the country for CAD functionaries, and provides matching grants to the States for State and Project level training for farmers and field functionaries.

Closing Remarks

The ever-increasing water requirements, and the dwindling availability, have already made water-related issues a matter of international concern. The issues are interrelated, and the dynamics of the future of mankind will be determined not by an individual issue but by the interaction of a multitude of issues related to integrated water resources development and management. The steady increase in population means the need for more food, energy and other raw materials. Augmentation of food and energy supplies requires prudent actions for appropriate and sustainable water resources development and management. Regulatory needs are to be identified suitably along with the evolution of suitable mechanisms for ensuring availability of water for all in acceptable quality and quantity. Effective and immediate steps are to be taken for the formation of River Basin and Water Regulatory Authorities with a will for implementation of the recommendations of the various commissions and tribunals.

Actions Needed

General

—Harness the untapped water wealth of the country.
—Realistic assessment of water resources of river basins, and the creation of a national information system and a database.
—Integrate water resources development and management.
—Add irrigation potential to meet the food and fiber requirements for ensuring food security.
—Reduce the gap between the irrigation potential created and utilized.
—Improve drainage for enhancing productivity.
—Create additional storage capacity through reservoirs and ensure optimal utilization.
—Maximize the hydropower development.
—Make safe drinking water available for all near their households.
—Safeguard the existing water resources from pollution and over-exploitation.
—Help communities revive the traditional water storage techniques and structures for rain water harvesting.
—Bring benefits of water resources development and augmentation of resources to scarce areas through transfer of water from surplus areas.
—Improve ecology by maintaining minimum water flows in rivers.
—Minimize the adverse environmental and social impact of water resources projects.
—Give due attention to health issues and water-related diseases.

—Mitigate miseries caused by water-related natural disasters through flood management and drought forecast.
—Develop navigation/inland water transport in the country.
—Ensure safety and serviceability of existing infrastructure and improve efficiency of irrigation water-use systems.
—Engage in active 'water diplomacy' for mutually beneficial use of water resources for overall economic development in the region.
—Arrest land erosion along river banks and sea coasts.
—Optimize the use of water as per agro-climatic conditions and drought proofing of arid areas.
—Stakeholders' and users' participation and informed decision making.
—Demand-side management through mass awareness.
—Dissemination and application of technology and research.

Policy Focus

—National Water Policy backed by Action Plan.
—Paradigm shift in emphasis towards improving the performance of existing infrastructure.
—Shift in strategy towards efficient management of flood plains; flood proofing including disaster preparedness and response planning, flood forecasting and flood insurance.
—Adopt a rational national resettlement and rehabilitation policy.
—Drought forecast, preparedness and mitigation measures.
—Consensus and agreement on transboundary waters with duty consciousness while claiming rights.

Administrative Initiatives

—Bringing all water-related subjects under one umbrella.
—Setting up of river basin organizations for integrated development and management of water resources.
—Re-orientation of research organizations.
—Creation of institutional arrangements with the requisite legal backing.
—Setting up of regulatory authority for rationalization of water rates.

New Legal Instruments

— Development of a National Water Code.
— Revision of the State Irrigation Acts to provide legal support to farmers' participation.
— Legislation for regulation and development of ground water on a sustainable basis.
— Legislation to protect and prevent encroachments of all water bodies.
— Legislative mechanism and process for settling water disputes.
— Legislation on dam safety.

Economic Instruments

— Ensuring the financial sustainability/maintainence of the existing facilities.
— Evolving improved economic analysis procedures.
— Encouraging adoption of insurance as a flood-management option.
— Linking subsidies and development assistance to financial and institutional reforms.
— Ensuring adequate allocation of financial outlays for the sector.
— Encouraging private-sector participation in development and management of water resources.
— Appropriate mechanisms for water pricing and recovery of water rates.

Social Change Initiatives

— Stakeholders' and users' participation including women's role in regulation and management of water resources.
— Creation of awareness on scarcity value, ecological and other issues of water sector.
— Integration of water uses and users.

Regulatory mechanisms are vital for integrated water resources development and management for poverty reduction, environmental sustenance and sustainable economic development.

Section III
Historical and Technological Perspective

6
Kaveri in its Historical Setting
S. Settar

Kaveri is one of the longest rivers in the Indian subcontinent, and is considered to be the holiest river in South India. Out of the 820 km of the river, 380 km are in Karnataka and 352 km in Tamil Nadu. Its coverage is 48 per cent in Karnataka, 44 per cent in Tamil Nadu, and 4 per cent each in Kerala and Pondicherry. In Karnataka, it is joined and fed by the rivers Harangi, Hemavati–Laksmanatirtha, Kabini, Suarnavati, Shimsha and Arkavati, while in Tamil Nadu it is joined by the rivers Bhavani, Noyil and Amaravati. It drains 42.25, 51.1, 3.55 and 0.2 per cent of land respectively in the South Indian states of Karnataka, Tamil Nadu, Pondicherry and Kerala. Depending on its availability, Karnataka, Tamil Nadu and Kerala contribute 53.8, 31.9 and 14.35 per cent of the water respectively into the river.

'Kaveri' of the Ancient Tamils

Kaveri is central to the social life of the states of Karnataka and Tamil Nadu. This has been true for at least a couple of millennia. The oldest body of literature in Tamil is called Sangam; by Tamil nationalists, it is dated between third century BC and third century AD, and by scholars it is dated in the first three centuries of the Common Era. In this literature, this river is called 'Kaveri' and occasionally, 'Ponni'. The Sangam Cholas — contemporaries of the Mauryas of Pataliputra in north India and the Chedis of Kalinga in the modern Indian state of Orissa in north-east India — ruled from Uraiyur, located on the banks of the Kaveri. One of the famous port-towns at the time was named Kaveri–Poompattinam. During this time as well as in the successive phases of history, a large number of devotees built temples on the banks of the Kaveri.

An estimated number of 20 Sangam poets, who hailed from the Kaveri area, have given an account of the role played by this river in the early life of the early Tamils. A poet belonging to the

Ahananuru — one of the two major genre played in the Sangam poetry — mentions that the river was always overflowing its banks. Another poet (*Aham*. 76) states that the beautiful river with its cool waters often broke the bunds built to regulate its flow. Its perennial character was such that a poet of *Purananuru*, another major genre in *Sangam* poetry, states that 'even if the Sun were to appear in the four quarters and the Venus or move in the southerly direction, the Kaveri would not stop feeding the universe' (*Puram*. 35). 'Kaveri suckled her children like the mother and fostered many lives', says yet another poet (*Puram*. 68).

According to the famous poet, Nakkirar, the land irrigated by the Kaveri in Tamil Nadu was so fertile that one *veli* of land produced a thousand *kalams* of paddy. Another poet says that a piece of land enough for an elephant to rest on yielded enough paddy to feed a family for the whole year. In another literary compendium, *Pattinapalai* (90), we are told that the fabulously-watered, cool and fertile fields, on the banks of the Kaveri, kept every one in the Chola kingdom happy. Here the peacocks danced joyfully, mistaking the drum beaten by the tiller in the field for the sound of the thunder squalls, and the sound made by the marching soldiers merged with the joyous outburst of happy swimmers in this river.

Not many *Sangam* poets discuss the origin of the Kaveri. The *Manimekalai*, one of the two earliest epics in Tamil, refers to the myth of Agastya. According to the myth, the Kaveri flowed out of the pot of Saint Agastya, when he accidentally tilted it. However, poets like Kausikanar were aware that the river took its birth in the Western Ghats, though its exact location appears to have been of little interest to him as well as to other poets of the time.

It is stated that a massive bund was built across this river for the first time by Karikala Chola in the second century AD. The existence of the bund is not a myth as it can be backed by historical records. Another important dam or *anekat* was built at Srirangam, almost a millennium-and-half later, by the Nayaks of Tanjavur. They also built step-ways that led to the river at this center, as well as at Mayavaram, Kumbakonam, Tiruvidi and Marudur. According to the *Kamandalu–vrittanta* of the *Kaveri Purana*, as narrated in the *Skandapurana*, it was Agastya who restrained the flow of the river by storing it in a pot he carried in his hand. This has been taken by some to suggest that this sage should be regarded as the earliest builder of a bund across the Kaveri!

Kaveri of the Ancient Kannadigas

In ancient Karnataka, this river was called Kaveri, Kaberi, Kamera, and sometimes, Sahyadri–Taneye — the last meaning the 'daughter of Sahyadri'. Two early kingdoms, called Punnatas and Gangas, built their capitals either on the banks of the Kaveri or in its proximity. The capital of Punnatas was at Kirtipura, and its hub lay between the rivers Kaveri and Kabini, while the capital of Gangas was at Talakadu, on the banks of Kaveri near Tirumakudalu. A sixth century record issued by the Punnatas reveals how gardens were developed in the fertile zone between these two rivers.

We do not know the role this river played in the imagery of the early Kannada folk poets, as all their compositions are lost. Though there was a general feeling that the upper Kaveri zone was not conducive for irrigational activities, nonetheless, repeated efforts seem to have been made to exploit its waters for cultivation of paddy, especially by the Kodavas, and other ancient communities residing in the Coorg region. In the *Deskat-Pat* of the Kodavas, reference is made to the cultivation of two types of paddy — white paddy in the upper region and red paddy in the lower region — '*paboliya-mendale-kengalele-kiggat*'. Some late ninth century records confirm paddy cultivation at Biliyur and other villages in the region.

The perception that the upper Kaveri region was not conducive for large-scale irrigation may have been due to the steep descent of the water, the undulating terrain of the region, and the relatively uncongenial soil conditions. Poets in Karnataka did, no doubt, preserve episodes that took place on the bank of the river, but most of these relate to wars fought between the ruling families of Karnataka (such as the early Chalukyas and the Rashtrakutas), and of Tamil Nadu (such as the Pallavas and the Cholas). Poets during the reign of the Chalukya kings, in the seventh and eighth centuries, describe how the massive army of Pulakesi, Vikramaditya and Vinayaditya checked the flow of the Kaveri between Uraiyur and Puhar, while marching against the Chola capital.

However, similar records of the period do not provide such details about activities around the Kaveri in Karnataka.

During the period of the Vijayanagara empire, the river saw some brisk activities regarding irrigation in the region occupied by the present Indian state of Karnataka. Half-a-dozen records, dated between 1467–1532, refer to the dams and bunds erected or re-erected at Sitapura, Srirangapattana, Triyambakapura, Byaladakere and other such centres, in the heartland of Mysore. This was continued in the late 17th and early 18th centuries, by the rulers of Mysore — the Chikkadevaraya of the Wodeyar family and in the later part of the 18th century by Tipu Sultan. Incidentally, one of the earliest references to the contested rights to the waters of the Kaveri is found in this period. When Chikkadevaraya Wodeyar completed building a dam, some Tamil chieftains of the Madurai region are stated to have hastened to Mysore and made an unsuccessful attempt to destroy it.

The first explicit reference to the Kaveri occurs in a Kannada literary work of the ninth century, called *Kavirajamarga;* incidentally, it is also the earliest surviving literary work in the Kannada language. It is attributed to the period of the reign of the Rashtrakuta king, Nrupatunga. The text states that the Kaveri (region) formed the southern boundary of the Kannada *janapada*, and in order to subdue the Gangas, who were ruling from Talakadu, the Rashtrakutas had to negotiate the impregnable river. Though the details of the history of this period are relatively speaking irrelevant for us here, it is interesting to note that the rugged terrain through which this river flowed in rapids seems to have made it difficult to harness its waters for agricultural purposes before the advent of modern dam-building technology.

Though the Kaveri is born in the Western Ghats, and flows hundreds of miles before entering into present-day Tamil Nadu, it was exploited more by the Tamils than by the Kannadigas. This is satirically observed by a 13th-century Kannadiga grammarian named Kesiraja, the author of *Sabdamanidarpana*. He wants to know from the Tigulas (the Tamils) whether they were borrowing the water (from Kannadigas), or appropriating it in the name of some past debt.

In the folk context, the river Kaveri was held sacred from time immemorial, but its far-reaching role is made evident in the literary works composed only during the last three centuries. In Karnataka, this seems to have begun in and after the 17th century, with the lead taken as much by the Tuluvas as by the Kannadigas. These mythical accounts have been called Kaveri *Puranas* and *Mahatmyas*.

Challenges and Responses during the Colonial Period

Kaveri offered some common challenges to the Tamils as well as to the Kannadigas, though more to the former, till the 19th century. The conspicuous ones among them centred round the flooding of this river. An *anekat* was constructed by Arthur Cotton during the years 1836–39, but floods continued to play havoc in the years 1858, 1896, 1906, 1911 and 1924. As many as 15 *anekat*s were built across this river, and another 21 across its tributaries in Karnataka. Similarly, besides the Grand, Upper and Lower *anekat*s, the Kaveri Bed Dam was constructed in Tamil Nadu in 1845 while Bhavani, Noyil and Amaravati received as many as 38 such constructions.

The introduction of several irrigation measures in the Mysore region seem to have been watched with some curiosity in the beginning of the 19th century in Tamil Nadu, but it only turned into a matter of considerable discomfort towards the end of the century. Restoration of old tanks and introduction of minor irrigation works between 1800 and 1810 by Dewan Poornaiah of the Kingdom of Mysore, and the constitution of the public works department in 1856 in the Mysore state were viewed with suspicion, and the Madras government formally expressed its concern on these developments in 1870. This brought the two states to the negotiating table at Ooty in the year 1890. The result of these negotiations was the Madras–Mysore Agreement, signed by both governments in the year 1892.

The 1892 Agreement put on record the schemes already completed, and those which were in progress in the Mysore region for the first time. This covered not only the works on the Kaveri but also on its tributaries. The agreement also laid down that no new projects should be undertaken by the Mysore government, and no attempts were to be made to restore the tanks which had been out of use for the previous three decades, without the concurrence of the Madras government. The agreement further stated that if any dispute on the developmental activities were to arise at any point of time in the future, it should be referred either to the Arbitration Commission or to the Government of India.

The 1892 Agreement did not satisfy either of the signatories because each felt that its age-old right had been ignored. Those in Tamil Nadu felt that the free flow of the Kaveri, which they were used to for millennia, had been curbed by this agreement; on the other hand, those in Mysore State felt that their right to make use of a river which took birth and flowed for a longer distance on their soil had been interfered with. Both these claims, no doubt, carried some weight. However, the issue was not one of traditional privileges, but of the compulsions of modern times. Water was certainly needed by the peasants of both the states, and this need could not be met satisfactorily without first readjusting the claims made through age-old practices. Nonetheless, the 1892 Agreement did bring to the surface the need to share a precious natural resource amicably. Yet, it did not provide any yardstick by which the dispute could be resolved to the satisfaction of all the parties concerned.

However, the 1892 Agreement did not come in the way of planning developmental activities either in Tamil Nadu or in Karnataka, but it made the two parties view each other with increasing suspicion. Between 1906 and 1910, the Madras as well as the Mysore governments proposed some major projects, in competition with each other. It was during these years that the Kannambadi Dam (also called Krishnaraja Sagar Dam), and the Kaveri Mettur Project (also called British–Kaveri or Madras–Kaveri Project), were proposed. When the Mysore government put forth its Kannambadi proposal, the Madras government emphatically argued that the Mysore project should not be cleared without first getting its Kaveri–Mettur project cleared.

Though the disputations continued for more than four years, very little came out of them. When the bilateral negotiations failed, the dispute — by then it had indeed become a dispute — was taken to the arbitration committee as per the 1892 Agreement. When the committee turned down the arguments of the Madras government, and upheld the case in favour of Mysore, the former made an appeal to the Government of India in 1915. When the appeal was turned down, the Government of Madras approached the secretary of state in 1919 and obtained a judgment in its favour.

The manner in which the case was pursued until the Government of Madras gained all it wanted showed that the 1892 Agreement was inadequate to solve the issues, and that negotiations outside the ambit of the agreement had to be made. Probably spurred by this, the bilateral negotiations were resumed by the two governments between 1920 and 1924. As a result of this, an agreement was reached regarding the construction of the Krishnaraja Sagar Dam. It is worth noting that the foundation stone of this dam had been formally laid as early as 1911, and that only after 13 years of negotiation it was realized by the Mysore government in 1924. No doubt the construction work was taken up soon after this, and the dam was completed in the year 1931. Not to be left behind, the work on the Mettur Dam was taken up by the Madras government in 1926, and was completed in 1934.

The 1924 Agreement produced tangible results because its main focus was on the construction of two specific projects viz. the Krishnaraja Sagar Dam and Mettur Dam, but incidentally, it also provided a framework for long-term irrigation development of the Kaveri basin in both the states. It was further agreed that the agreement would be reviewed after 50 years.

Challenges and Responses during the Post-Colonial Period

It was not forseen in 1924 as to what would happen if this agreement was not renewed at the end of the fiftieth year. When this did happen in 1974, it led the two States to interpret the 1924 Agreement in accordance with what suited each best. The Tamil Nadu government held that the 1924 Agreement being the final settlement was binding on both the governments even after the expiry period, while the Karnataka government felt that since it was not renewed, the agreement had become redundant after 1974.

The developmental work, however continued in both the states, notwithstanding the ongoing debate on the sharing of the waters of the Kaveri. The irrigation facilities available for the peasants in Mysore State in 1900 was sufficient for cultivating only 1.1 lakh acres; this increased to 4.4 lakh acres in 1971. This was made possible by the irrigation work taken up on Kabani (1959), Harangi (1964), Suarnavati (1965), Hemavati (1968), Varuna (1979) and Yagachi (1983). The developmental work in Tamil Nadu was no less remarkable. Here, the irrigated lands, before the completion of the Mettur Dam, numbered 14.4 lakh acres and this was increased to 45 lakhs after building the dam. The major works by the Tamil Nadu government between 1951 and 1961 comprised the Lower Bhavani Project, the Mettur Canal Work, and the Amaravati Project. These were followed by the New Kalatalai High Level Canal and the Pullamalai Canal.

The sharing of waters of the Kaveri became a major issue of conflict between the Tamil Nadu and Karnataka governments during the latter-half of the 20th century, and the early years of the 21st century. Kerala and Pondicherry also joined by putting up their claims in 1970 and 1978 respectively. Though the 1892 and 1924 agreements, made during British rule, did surface now and then, they were increasingly marginalized by the newly formed linguistic states.

Protection and promotion of the interests of peasants of each linguistic state became the foremost responsibility of those in power in the four states of South India. Also the issue did not remain the exclusive concern of the state governments, as it was during the British rule. Organized groups of peasants began to step in to make their demands and presence felt. Under these changed circumstances, the creation of new platforms for negotiations became inevitable. This led to the organization of a series of bilateral and multilateral meetings. Between 1968 and 1990, 22 state ministerial-level meetings, 21 meetings under the chairmanship of the union minister of irrigation, five bilateral sessions between Tamil Nadu and Karnataka took place. Many of these were held in the seventies, but the number of these meetings became scarce after the eighties.

During the seventies, successive governments in Tamil Nadu began to demand the constitution of a tribunal under the Inter-State Water Disputes Act of 1956, and urged for the tribunal enquiry, which took place in 1975, 1979 and 1986. In the meantime, in 1971, Tamil Nadu and Kerala appealed to the Supreme Court. Partly in response to this, the Kaveri Fact Finding Committee was constituted in the year 1972. In 1986 the farmers of Tanjavur also took initiative to move the Supreme Court. However, the stalemate was continued and both the governments found very little scope for a meeting ground. In 1988 the Supreme Court directed the governments of Tamil Nadu and Karnataka to resume bilateral meetings, and two years later, in 1990, the Kaveri Water Dispute Tribunal was established on its direction. Recently, in 2007, the Tribunal gave its award that left the Government of Karnataka dissatisfied. Following the award, the latter filed a review petition before the Tribunal. The dispute continues to fester, and innovative mechanisms to resolve it need to be evolved. But this can only happen if we locate this dispute within its proper historical context.

7
The Cauvery Tribunal Award from a Hydrology Perspective

Rama Prasad

Introduction

The river Cauvery rises in the Western Ghats near Bhagamandala in Kodagu district in Karnataka and flows through Tamil Nadu into the Bay of Bengal. Its tributaries include Hemavathi, Harangi, Kabini, Bhavani, etc., some rising in Karnataka and others in Kerala and Tamil Nadu. Its delta includes Karaikal, a part of Puducherry (formerly Pondicherry). Thus, there are four states in the Cauvery Basin (Map 7.1). The Cauvery water dispute arose in the 19th century when the then state of Madras (now Tamil Nadu) objected to some irrigation works proposed to be undertaken by Mysore (now Karnataka), and led to an agree-ment being signed by Madras and Mysore in 1924 that inter alia prescribed how much new area could be irrigated in each state. A history of the dispute is beyond the scope of the present article.

The Cauvery Water Dispute Tribunal was constituted by the Government of India in 1990. After more than 16 years of work, the tribunal gave the Award. Highlights of the Award are:

1. The 50 per cent dependable[1] annual yield (i.e., quantity of water) in the Cauvery at the Lower Coleroon Anicut (LCA) is 740 TMC.[2]
2. Out of the above yield, Karnataka can utilize 270 TMC,[3] Kerala 30 TMC (out of which 21 TMC is in Kabini basin), Tamil Nadu 419 TMC and Pondicherry 7 TMC.
3. If the yield in any year is less than 740 TMC, the above shares are reduced proportionately.
4. Karnataka should release 192 TMC at its border with Tamil Nadu.
5. A regulatory authority should be set up to implement the Award.

Map 7.1: Cauvery River Basin, Irrigation and Hydro-electric Projects

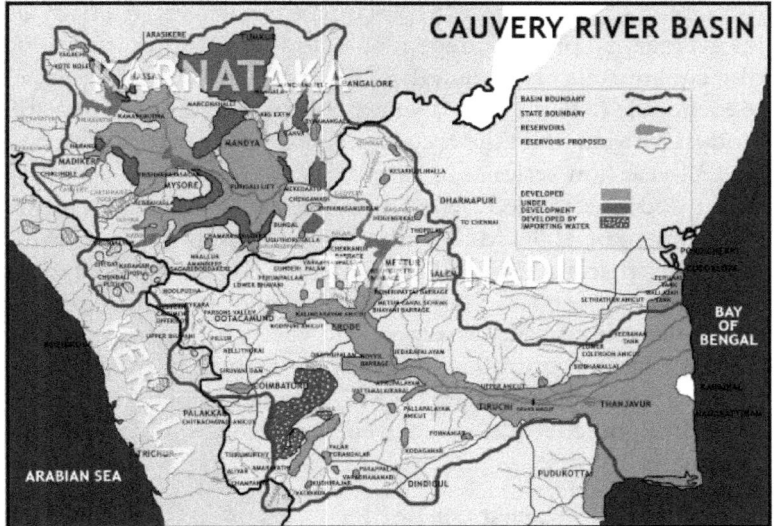

Source: Water Resources Development Organisation, Government of Karnataka. Adapted by Mr Raja P.K.

Yield

As part of the efforts for a negotiated settlement of the Cauvery dispute, the Government of India had constituted a Cauvery Fact Finding Committee (CFFC) in 1972 with a mandate to submit its report in three months. The CFFC collected data from the basin states and stated in its report that the 50 per cent dependable annual flow in the river at Lower Coleroon Anicut (LCA) is 740 TMC and the 75 per cent dependable flow is 670 TMC, on the basis of data from 1934–35 to 1971–72.[4] Although the flow at LCA was measured from 1900 onward, the CFFC considered only the data from 1934–35 up to 1971–72. The Tribunal considered the CFFC figure of 740 TMC for allocation, ignoring data earlier than 1934 or later than 1972, which is something no hydrologist would do. Curiously, the CFFC did not give the annual flow series in its report, and from the available data it is not possible to reproduce the CFFC's flow figures (incidentally, the methodology of yield calculation given by CFFC in its report leads to some physically meaningless number, and not the annual flow).

The reason hydrologists insist on the longest possible record is that the standard error (SE) of quantities derived from it (such as average, 50 per cent dependable flow, etc.) varies inversely as the square root of the length of record. The SE of 50 per cent dependable flow is 1.253 $\sigma/\sqrt{n}$, where σ is the standard deviation and n is the length of the record (i.e., number of years of data). If the 39-year flow series at LCA from 1934–35 to 1971–72 derived from documents filed before the Tribunal is considered, its σ turns out to be 165 TMC. With $n = 39$, the SE is 33 TMC, which is greater than the 30 TMC allocated to Kerala! Ideally, the SE should have been less than the smallest allocation unit (which is 4 TMC for inevitable wastages). However, it can be argued that wastage is not controlled and may be zero in particular years. If the SE is to be less than 7 TMC (the next smallest allocation unit, to Pondicherry) the length of the record should have been at least 871 years!

Release at the Karnataka Border

The Tribunal has ordered that Karnataka should release 192 TMC at Biligundlu (situated on the border with Tamil Nadu, *vide* Table 7.1) in a year. Having ignored all post-1972 flow data, it had to use the numbers contained in the CFFC Report to arrive at the release as follows:

Table 7.I: Water Budget above Mettur

Yield at Mettur	508 TMC
Less yield between Karnataka border and Mettur	–25 TMC
Yield at Karnataka border	483 TMC
Less utilization allowed for Kerala in Kabini	–21 TMC
Yield at border after Kerala utilization	462 TMC
Less utilization in Karnataka	–270 TMC
Balance available at border, to be released to Tamil Nadu	192 TMC

Source: Prepared by author, based on data from the Final Report of the Cauvery Water Disputes Tribunal (February 2005).

The annual release is broken into monthly quantities (in TMC) as in Table 7.2, but the hydrological or agro-meteorological basis for the break-up has not been revealed. In particular, it is not clear whether there is enough water to meet the monthly releases ordered after Karnataka and Kerala have drawn their requirements.

Table 7.2: Monthly Break-up of Release at Karnataka–Tamil Nadu Border (in TMC)

Jun	Jul	Aug	Sep	Oct	Nov	Dec	Jan	Feb	Mar	Apr	May
10	34	50	40	22	15	8	3	2.5	2.5	2.5	2.5

Source: Final Report of the Cauvery Water Disputes Tribunal (February 2005).

Return Flow

Return Flow from Evapotranspiration Considerations

The consumptive use (evapotranspiration)[5] of most non-perennial crops is of the order of 600 mm. This figure tends to be uniform across most crops of similar duration since evaporation and transpiration are determined largely by solar radiation, with small variations introduced by other factors. It is therefore assumed that evapotranspiration in the irrigated areas is 600 mm. Any water supplied to the crops in excess of this amount infiltrates into the ground and raises the water table. Once the water table intersects the river bed, ground water flows into the river. Such flow is called irrigation return flow. Return flow occurs from water supplied for domestic and industrial uses also. The Tribunal has taken the return flow from domestic supply as 80 per cent and from industrial supply as 97.5 per cent while making water allocations. However, no return flow from irrigation is considered. Table 7.3 gives the return flow from irrigation based on the 600 mm evapotranspiration criterion (there is a small area of perennial crop for which the evapotranspiration is higher than 600 mm, but this has been ignored for the sake of simplicity. Also, during the initial

Table 7.3: Return Flow on the Basis of Evapotranspiration

	Karnataka	Tamil Nadu
Area irrigated, lakh acres	18.85	24.71
Water allocated by Tribunal, TMC	251.00	391.00
Depth of water, mm*	932.00	1107.00
Evapotranspiration, mm	600.00	600.00
Return flow, mm£	332.00	507.00
Return flow, TMC#	89.00	179.00

* (Area irrigated ÷ Water allocated), converted to mm units.
£ (Depth of water — Evapotranspiration).
(Return flow in mm × Area irrigated), converted to TMC units.

Source: Prepared by author, based on data from the Final Report of the Cauvery Water Disputes Tribunal (February 2005).

days of irrigation, ground water will be below the river bed and will not flow into the river. This has also been ignored and it is assumed that all the infiltrated water becomes return flow).

These return flows are spread out over the total irrigated area. Return flow from an upper area will not only be available for reuse from a downstream structure (such as a dam or anicut), but will also get included in the flow measured downstream. It should be noted that the allocations are on the basis of flows from 1934–35 to 1971–72, when irrigation in Karnataka was not developed to the extent we see today, and in particular, there was not much return flow above Krishnarajasagara (KRS) Dam. For example, part of the return flow from the KRS command thus got included in the measurement at Mettur, and, in addition, that from the command of Lower Bhavani Dam and river channels in the measurement at Grand Anicut. It needs a complex analysis to determine which part of the return flow has been reused and which part got into the 740 TMC. This problem does not exist downstream of the last measuring point. Grand Anicut being the last measuring point on the main Cauvery, the return flow from the delta (old and new) did not get included in the 740 TMC. So also the return flow from the LCA command, LCA being the last measuring point on the Coleroon. Thus, if the tribunal order were strictly implemented, the return flow from the delta and LCA command would flow into the sea. The allocation for the 10.27 lakh acres in the old delta is 150 TMC, and consumptive use at 600 mm amounts to 88 TMC. The balance of 62 TMC would flow to the sea. Similarly, 23 TMC out of the 46.5 allocated for 2.76 lakh acres in new delta and 7.5 TMC out of 19.5 allocated to the LCA command of 1.4 lakh acres would flow to the sea, all totalling 92.5 TMC.

Return Flow from Irrigation Efficiency Considerations

The Tribunal has assumed an irrigation efficiency[6] of 65 per cent for most of the irrigated area. This means that 35 per cent of the water drawn from the river/reservoir is 'lost' , i.e., will infiltrate into the ground and become return flow. On this basis, the return flow in Karnataka would be 88 TMC (35 per cent of 251) and in Tamil Nadu, 137 TMC (35 per cent of 391). Taking the last measuring points as before, 52 TMC (35 per cent of 150) from the old delta, 16 TMC from the new delta (35 per cent of 46.5) and 7 TMC from the LCA command (35 per cent of 19.5) would flow into the sea, totalling 75 TMC. These figures are similar in magnitude to the ones obtained above from consumptive use (evapotranspiration) consideration.

Percolation 'Losses' in Paddy Irrigations

The tribunal has assumed that the standing water in paddy fields[7] percolates into the ground at the rate of 3 mm/day in Karnataka and 2 mm/day in the old delta, 2.5 mm/day in the new delta and 3 mm/day in the non-delta areas in Tamil Nadu. It has included this percolation in computing water requirement for paddy, and treated it as a loss. This water infiltrates into the ground, raises the water table and seeps into the river downstream. The percolation 'loss' also therefore behaves like return flow. In Karnataka and non-delta region of Tamil Nadu, irrigation canals are taken from reservoirs or anicuts and run on higher ground above the river. The percolated water as well as the return flow reaching the river is therefore not available for irrigation and is 'lost' unless there is an irrigation structure downstream from where it can be used. However, in the delta, the river divides into numerous branches, and these branches themselves act as the main canals. Return flow from the water drawn from these branches rejoins the branches downstream. There that water is again available for use, and this fact distinguishes the delta hydrologically from the non-Delta areas. Hence, water percolating at 2 mm/day in the upstream reaches of the delta would join the river branches in the downstream reaches and would immediately be available for meeting percolation loss in that reach, and so on. It may take two or three days for all the percolated water in a given reach to come back to the river, since it has to flow through the soil (and therefore with velocity lower than surface water). Hence, all the percolated water on the first day would be available for meeting the percolation requirement on the fourth day, water percolating on the second day will be available on the fifth day, and so on. Table 7.4 illustrates this point. Thus, it would be necessary to allocate water for percolation only for the first 3 days.

Instead, the tribunal has provided for percolation for 115 days on 9.5 lakh acres of Samba[8] plus 160 days on another 3.6 lakh

Table 7.4: Percolated Water

Day	Initial flow in river	Consumptive use requirement	Percolation requirement	Release into river	Total flow in the river	Diversion to field	Return to river
	mm	mm	mm	mm	mm	mm	mm
1	0	ETcrop	2	ETcrop+2	ETcrop+2	ETcrop+2	0
2	0	ETcrop	2	ETcrop+2	ETcrop+2	ETcrop+2	0
3	0	ETcrop	2	ETcrop+2	ETcrop+2	ETcrop+2	2 from Day 1
4	2	ETcrop	2	ETcrop	ETcrop+2	ETcrop+2	2 from Day 2
5	2	ETcrop	2	ETcrop	ETcrop+2	ETcrop+2	2 from Day 3
6	2	ETcrop	2	ETcrop	ETcrop+2	ETcrop+2	2 from Day 4

............and so on till the last day of season.

Source: Prepared by author, based on data from the Final Report of the Cauvery Water Disputes Tribunal (February 2005).

acres of Kuruvai and Thaladi[9] in the old and new delta. The excess percolation provision (over and above 3 days) amounts to 49 TMC. However, 35 per cent of this has already been included in the return flow in the previous section. Hence, the additional amount of water flowing to the sea on account of the excess is 32 TMC. Thus, at least 107 TMC of surface water (75 + 32) has been left unallocated.

Ground Water

Ground water in the delta was a contentious issue before the tribunal. Extensive studies made by the UNDP and others have put the minimum availability of ground water in the delta at about 30 TMC. The tribunal reduced it to 20 TMC without any hydrological basis, and surprisingly, ignored it altogether in the final award. In contrast, it assumed that half the drinking water requirements in the basin states would be met from ground water, though there were no studies to support this assumption.

Implementation Problems

Implementation of the award poses several problems, some of which are discussed below:

Utilization and Release Problem

Unlike the Narmada and Krishna tribunals (which did not order any release by upper riparian states), the Cauvery tribunal has ordered allocations to party states as well as release by Karnataka into Tamil Nadu. Karnataka is to release 192 TMC at the border with Tamil Nadu if the yield in the river (at LCA) is 740 TMC. Now, as assumed in effect by the tribunal (Table 7.2), the yield at the Karnataka–Tamil Nadu border is 483 TMC, and below the border 257 TMC. Above the border, most of the water comes in the months of June–October, and below the border, in the months of August–January. Yields in the two parts are not strongly correlated. If in a particular year, yield above the border is, say, 450 TMC and below the border 290 TMC (the total being 740 TMC), Karnataka cannot utilize 270 TMC and also release 192 TMC (it should be kept in mind that 21 out of the 483 TMC is utilized in Kerala). There will thus be a problem if Karnataka insists on utilizing 270 TMC and Tamil Nadu insists on a release of 192 TMC.

Yield Determination Problem

If the yield in a particular year ('distress year') is less than 740 TMC, the allocations are to be reduced proportionately. Now, 'yield' refers to the flow which would have occurred at LCA had there been no utilization upstream (counting the flow into Cauvery and Vennar at Grand Anicut as utilizations, as done by the Tamil Nadu Irrigation Department). In practice, since there are utilizations, the yield is calculated as the measured flow at LCA *plus* all the utilizations upstream. It is here that the problem arises. There are thousands of tanks and dozens of anicuts where the utilization, totalling more than 200 TMC, is not measured at all, but estimated based on the respective command areas. This may be good enough to arrive at a rough estimate of the average yield, but is not suitable for determining the share of basin states in a distress year as ordered by the tribunal. Strictly speaking, it is not possible to say what the yield in a given year is. The basis of allocation being the 50 per cent dependable flow, five years out of ten will be distress years. Since the actual yield cannot be determined accurately, there will be disputes in the regulatory body every alternate year or even every year about whether there is distress or not.

Operation Problem

Water utilization has to start in June, but whether the yield is less than 740 TMC (even if we ignore the accuracy issue) would be known only at the end of the following May. Even rough indications whether there is likely to be a substantial deficit in the yield would be available only in October, and by that time the crop season would be nearing the end. Indications of relatively small deficits of the order of 30–40 TMC will come even later. The tribunal's order with regard to reservoir operation only makes matters worse: *the reservoirs are to be operated on a 10-day basis and any deficit in the 10-day flow has to be shared by all states.* There is no benchmark to compare and say whether yield in a particular 10-day period is in deficit or excess. We have only the one number 740, and that is for one year.

Measurement Error Freezing Problem

The utilizations from the Cauvery and its tributaries are measured at several reservoir and anicut sites by the states. In a very large number of cases, the measurement is by applying discharge formulae with uncalibrated coefficients or reading from unrevised stage-discharge tables. These recorded utilization values would therefore have unknown errors compared to true values. An obvious example is provided by the comparison of flow recorded at Mettur (using uncalibrated formulae) with that at Biligundlu (using current meter), situated a few kilometres upstream of Mettur. Many small and large streams join the Cauvery downstream of Biligundlu, and this addition of water should have been reflected in a higher flow at Mettur (no water is diverted from the main river in between). But the flow recorded at Mettur is *less* than that at Biligundlu (Figure 7.1) in most years. Errors pertaining to structures in existence during the period from 1934–35 to 1971–72 are built into the annual yield of 740 TMC. Any improvement or deterioration in the accuracy of measurement at these structures would lead to yield figures that cannot be compared with the benchmark 740 TMC. Thus, errors in measurement have to be maintained at 1934–35 to 1971–72 levels in order to decide whether there is a deficit in yield.

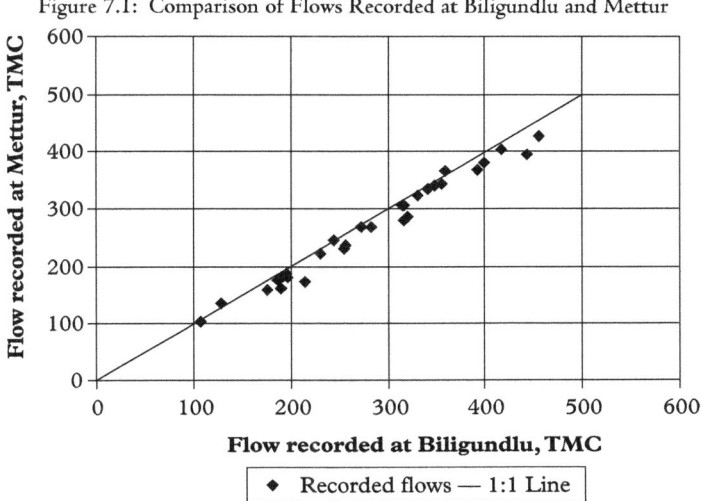

Figure 7.1: Comparison of Flows Recorded at Biligundlu and Mettur

Source: Prepared by author, based on data from the Water Resources Development Organisation, Government of Karnataka.

Freedom-Obligation Conflict Problem

The tribunal has given the freedom to each state to utilize its allocation in any manner and anywhere it wants. Thus, although Kerala's allocation of 30 TMC has been calculated on the basis of 21 TMC in Kabini and 9 TMC in Bhavani and Pambar, it is free to utilize all 30 TMC in the Kabini basin. If Karnataka utilizes its 270 TMC, there will be only 183 TMC left at Biligundlu in a normal year, and 192 TMC cannot be released there.

Conclusion

The Cauvery tribunal Award contains too many hydrological loose ends to be properly implemented. It is not possible to determine the yield to a reasonable accuracy in any particular year, nor is it possible to decide whether there is distress or not, especially before the end of the irrigation season. There are also various implementation problems, which may not be amenable to solution at the regulatory authority level. A substantial quantity of water appears to have been left unallocated due to non-consideration of irrigation return flow.

Notes

1. The term 'dependability' though widely used in Indian irrigation engineering practice, is not found in hydrology literature, where it is known as probability of exceedence. The 50 per cent dependable quantity is called the median in statistics books. The river gets its water from rainfall, which can be roughly partitioned into run-off, evapotranspiration and ground water. Run-off is the flow in the river, evapotranspiration is the evaporation of soil moisture by vegetation and soil, and the remaining rainwater percolating deep into the ground is ground water.
2. 'TMC' stands for a thousand million cubic feet, one TMC being equal to 28.32 million cubic metres. The unit TMC is used throughout in this article in preference to the metric unit since most of the documents relevant to the dispute use that unit.
3. The river Cauvery branches into two at the Upper Anicut (upstream end of Srirangam island). The northern branch is called Coleroon (Kollidam) and the southern branch retains the name Cauvery. The Coleroon continues as a separate river beyond Grand Anicut (downstream end of Srirangam island). A weir known as Lower Coleroon Anicut (LCA) built on the Coleroon (Figure 7.1) is the last point where the flow in the Cauvery system is measured.
4. Water Year 1934–35 means the twelve months from June 1934–May 1935 (monsoon starts in June).
5. Evapotranspiration is the combined evaporation from vegetation (transpiration) and soil (evaporation).
6. Part of the water let out into a canal from a reservoir or anicut seeps into the ground from the sides and bed of the canal. Part of the water applied to the field also seeps into the ground. The remaining water is what is actually used by the plants (consumptive use). The ratio of consumptive use to the water let into the canal is called irrigation efficiency.
7. It is the usual practice to allow water to stand to a certain depth in a paddy field for most of the growing season, as this is found to increase the crop yield.
8. Samba is a long duration rice variety grown in most of the Cauvery delta. It is planted in August so that it receives support from the north-east monsoon rainfall, which is the dominant rainy season in the delta.
9. Kuruvai is a short-duration rice variety grown in a small part of the delta. It is planted in June, and depends partly on ground water and partly on the Cauvery water, as rainfall support is small. After it is harvested, Thaladi, another short-duration rice, is planted on the same field.

Section IV
Negotiated Approaches and Alternative Paradigms

8
Evolving a Negotiated Approach to Sharing of Transboundary Rivers

Vijay Paranjpye

Introduction

All major rivers in India are currently embroiled in conflicts relating to sharing of waters between contending states which do not appear to be seriously interested in finding solutions that are acceptable and beneficial to all concerned. This perception is based on the evidence available on records related to the seven disputes that have been referred for arbitration under the Inter-State River Water Disputes Act, 1956. Even in cases where awards have been declared after decades of long drawn-out arbitration, the legislators of state governments concerned and the respective bureaucrats and technocrats have openly disregarded the tribunal awards, and made statements that smack of contempt for legal procedures and democratic institutions. Subsequently, the press and society-at-large, often incited by the state authorities themselves, join in the unholy pandemonium. Unfortunately, many well-meaning experts on water management have blamed the inadequacies in the legal and policy framework for this sad state of affairs, instead of realising that the state governments have been responsible for completely undermining the legal instruments and making a mockery of the judgement and awards.

Why Has Legal Arbitration Failed?

The major contention of this article is that the minimum level of societal awareness and involvement in the democratic process necessary for making the relevant policies and laws effective does not exist as yet in India. Therefore, it is necessary to work on these prerequisites in a conscious and purposive manner, rather than finding faults with the statutory processes. Recent experiences suggest that creating the right societal attitudes is time-consuming and requires inputs from civil society organizations and motivated facilitators and negotiators. However, considering the decades lost in statutory arbitration, spending time and efforts in creating the required knowledge base and building up appropriate social perceptions and democratic institutions would be certainly worthwhile. Such a process would require:

1. The creation of the necessary database and framework for analyzing the physical, ecological, social and economic aspects of river basin.
2. Information dissemination, well-informed public debates through mass media or multi-stakeholder platforms.
3. A culture of democratic participation capable of understanding and participating in public discussions on contentious issues related to the sharing of waters of inter-state rivers.
4. The next logical step would be to bring about negotiations between stakeholders and government authorities of two or more states on a common platform for arriving at solutions through discussions on possible trade-offs between alternative solutions. At this stage, the prerequisites necessary for resolving contentious issues are: administrative transparency, and a sense of mutual accountability and trust, and respect and responsibility among the stakeholders, the authorities and the negotiators. It is only when the society has gone through these stages and yet failed to arrive at water-sharing solutions, that matter be taken to inter-state water tribunals for statutory arbitration. In such cases, when an award is declared, there is greater likelihood of it being accepted and implemented, since all other options have failed and the society is aware of all consequences of abiding (or not abiding) by the judicial award.

The recent examples in Punjab, Maharashtra, Karnataka and Tamil Nadu indicate that this process was not followed and the matter was referred directly to the Inter-State River Water Dispute Tribunals by the state governments, without taking the people into confidence. In the absence of a well-informed constituency, political manipulation led to a loss for all concerned.

The most disturbing case was that of the Sutlej–Yamuna Link Canal, wherein the non-completion of the canal in Punjab resulted in both Haryana and Punjab taking the matter to the Honourable Supreme Court. The Supreme Court directed the central government to carry out the action plan for completion of the canal, under a judgment passed on 4 June 2004. Accordingly, the central government started taking action. However, The Punjab legislature on 12 July 2004 enacted, 'The Punjab Termination of Agreement Act, 2004', thereby terminating its own obligations under the earlier agreements. The rejection of the Supreme Court decision by the Punjab government was a clear case of contempt for the legal system and a collapse of democratic procedures.

The other example is of the recent violence that erupted after the declaration of the Cauvery Award. Ironically, all the major contending parties, namely Karnataka, Tamil Nadu, Kerala and Pondicherry appear to have revolted against the Award! Considering the long history of the conflict and the 17-year-long judicial process this was a tragic result.

The third recent example is that of the Babhali Dam in Maharashtra, on the Godavari river, which led to violence on both sides of the Maharashtra–Andhra Pradesh border. Interestingly, in the case of the Godavari Award, negotiated agreements had already been reached between the two state governments, and the Award had only reiterated those agreements. However, the paradoxical reaction of the people was due to the lack of transparency during the process of negotiation and arbitration, and exclusion of stakeholders from all discussions.

The events and media reports suggest that when the negotiated settlements had been reached between the governments, the details regarding the agreement were never communicated or revealed publicly. Consequently, even though the decision to construct the Babhali Dam was within the framework of the earlier agreement, the people in Andhra Pradesh were misled into believing that the dam would harm their interests. This led to protests on both sides. Such behaviour occurs because after being elected, legislators do not feel that they are accountable or responsible to the electorate in such matters. Till now seven major tribunals have been established so far, and practically in all cases the final outcome has been frustrating because people have not been involved or taken into confidence and are therefore apprehensive or resentful of the decisions.

The Lack of Integrative Thinking in River-Basin Development and Management

Another important reason behind the failure of inter-state arbitration is the lack of integrative thinking among all parties concerned. In India, the administrative agencies in charge of development activities are so compartmentalized, that they can rarely interact with each other for getting benefits of synergetic interaction, or for arriving at well-negotiated optimal solutions. In fact, the corporate enterprises and civil society organizations also often fall into the same trap. The result is, that in spite of possessing the requisite technological competence, financial resources and administrative/managerial skills related to natural resources development, the results are appallingly unsatisfactory. While the political, administrative and social units of planning and development like the states, districts, talukas, etc., are based on linguistic or other administrative considerations, the water-sharing issues are based on the physical catchment boundaries or watersheds and smaller tributaries. There is a clear mismatch between ecological boundaries and the political or administrative divisions which inhibit basin thinking.

This situation is particularly evident in the case of water resource development, which requires the integration of agencies dealing with a) soil and water conservation, b) minor, medium and major irrigation works, c) urban and rural water supply *and* sanitation, d) flood-control, e) wetlands and lake conservation, f) forestry, g) stream and river-bund protection, h) navigation, i) hydropower, j) ground water, and k) inland fisheries, etc.

About Principles, Processes and Parameters

Sharing of water by riparian states along a river basin is a sociopolitical phenomenon, and it involves notions of justice and fairplay in addition to emotional feelings and pride. These notions, in turn, are highly subjective and relative in nature. To respond to these, one needs broader principles and social processes which build institutions and mechanisms, which involve all stakeholders whose understanding of river-basin evolves and matures along with the institutions.

On the other hand by parameters one generally understands quantifiable entities, which can be numerically quantified and measured with a certain degree of exactitude. Parameters are objectively verifiable and can be validated scientifically and are not affected by emotions like sympathy, magnanimity and (mis) trust. Although one would like to develop such parameters for resolving conflicts about water sharing, identifying the ones which will command social acceptance are difficult to come by.

The article at hand has therefore tried to restrict itself to 'participatory processes' and 'principles' which could enable effective and timely resolution of water conflicts.

Civil Society Must Step in to Achieve Integration

Since there does not appear to be a set of formal procedures for integrating all these elements, other ways and means need to be devised and instrumentalities established for initiating discussions, sharing of knowledge, entering into dialogue, arbitration and finally reaching the stage of conflict resolution. There is also a need for establishing a forum, which not only permits divergent views to come to the negotiation table, but also advocate for social and political legitimacy to the process. This alone would enable the democratic implementation of solutions, agreements, and awards thus derived.

It is our contention in this article that voluntary agencies and other civil society organizations are in a position to be flexible and innovative enough to operate in such complex situations requiring co-ordination and negotiation within the various government departments and local stakeholders. Therefore, for achieving the goals of integrated water resource management in general and integrated river basin management in particular, states will have to create a knowledge base, publicize it and follow it up with informed public discussion, and multi-stake holder negotiations. Conflict resolution proceeded by such a process is the most effective and desirable way of developing a societal consensus on transboundary issues.

The Gomukh Trust has worked in this area for a decade and the experiences accumulated indicate that it is in a position to integrate several, if not all concerns into its approach, and that the results over the past decade have been very promising. Gomukh at times has had to take recourse to the path of confrontation and litigation, but is convinced that persistent and sustained negotiation and dialogue is the best option for conflict resolution.

A review of the inter-state conflicts for water sharing suggests that the problem of allocation of water between states is a contentious and vexing issue, mainly because the deeper underlying conflicts which are regional or local, have remained unattended and unresolved. For example, the resolution of conflict between upstream and downstream irrigators, between urban/industrial water users and rural water users, conflicts due to resource ownership and entitlement, extreme inequity in distribution, conflict between service providers and water users, conflicts arising due to displacement of people for water infrastructure development projects, etc., are matters of immediate concern to the community. Addressing these and resolving them leads to awareness about the broader river-basin perspective within which these local issues are located. Interactions also lead to a better understanding and appreciation of other legitimate viewpoints.

A large number of intrinsic conflicts arise because water has been treated as a commodity to be dealt with by different departments/sectors for satisfying various water demands. In actual fact, all of them are inter-related and inter-connected. Unless they are treated as inalienable parts of the river basin ecosystem, they would inevitably lead to conflicts and contradictions. Treated in isolation, the actions of one sector are often detrimental to others and vice-versa. Therefore, in this section, we advocate that instead of dealing symptomatically with conflicts, it would be desirable to adopt a negotiated approach to water management in general and make conflict resolution a part of this broader approach.

The negotiated approach to integrated river basin management (NA-IRBM) asserts that local communities have the potential of managing natural resources not only in their immediate vicinity, but also of upscaling the vision to cover the entire river basin. A river basin has been taken as the reference ecosystem in this case because the NA-IRBM evolved mainly as a response to growing dissatisfaction amongst community and Civil Society Organisations (CSOs) to conventional, top-down water management approach.

The approach is not restricted only to applying the 'subsidiarity principle', where the communities take part in decision-making restricted to their own village or sub-catchment, but it 'calls for the reverse, allowing local actors to develop basin management strategies specific to their local context, which are then incorporated in the larger basin management plan. This allows their knowledge to influence regional and national decisions, ultimately resulting in a truly bottom-up process of policy development and management' (Hirsch and Paranjpye 2005).

Another important facet of negotiation is managing the 'tradeoffs' that arise from the interactions between ecosystems and livelihood. Conventional management and policy has little to contribute to the 'issue of "winners" and "losers" associated with ecosystem changes, and in particular the impact of ecosystem changes on marginalized communities, which has not been adequately taken into account in management decisions' (United Nations 2005). Through the negotiated approach, community platforms, often based on traditional institutions, are nurtured and strengthened to address these issues. Capacity building and guidance from external organizations forms an important part of the process to ensure strong representation from weaker stakeholders like families below the poverty line, women and especially, environment. Thus, the approach is considerably different from decision-making based on the majority opinion.

Negotiating for Livelihoods and Health of Ecosystems

The societal dissatisfaction about the reservation of 10 per cent of water for ensuring 'environmental flows' by the recent Cauvery Award (2007) was obviously due to a lack of understanding about the positive relationship between livelihoods and maintenance of healthy riverine ecosystems. People felt that while there was a scarcity of water for human consumption and agricultural needs, water was instead being reserved for the environment.

Such erroneous perceptions can be changed only through information and awareness about the linkages between the necessity of water for ecosystems and its relationship to human life. It must be emphasized that the protection of ecosystems and poverty alleviation are interrelated. In most of the developing countries, the two are closely interlinked. But, due to unsustainable anthropogenic pressures, the carrying capacity of ecosystems is severely jeopardized, thus affecting ecosystem goods and services on which the rural poor depend directly. Attempts at focusing on one of the aspects, without working on the other have proven to be unsustainable.

Anatomy of Water Conflicts

All conflicts can be broadly put in the following categories.

> (a) **Quantity**: The quantity of water being finite in any river basin disputes arise when one party happens to appropriate a disproportionately larger share of water and others are aggrieved because they claim an equal share of the resource. The lower riparian communities routinely blame the upstream users for shortages in their own areas, and conflict situation emerges.
> **The Zero Sum Game**: The most intractable and irrational cause of water conflict arises from when different contending parties (for instance, the states of Kerala, Karnataka and Tamil Nadu in the case of the Cauvery dispute) demand shares of water which together add up to a volume greater than the annual average flow of water available for use.

(*b*) **Quality:** Using the same logic the downstream settlements blame the upstream users for polluting the water, thereby making it either useless for certain purposes or too expensive to use after treatment and purification. This leads to exclusion of affected communities. It is usually the industries and metro cities, which are blamed for using excessive amounts and polluting the water in river-regimes downstream.

(*c*) **Historically Appropriated Right:** The appropriation of water during times of relative abundance, e.g., 20 per cent of Krishna waters were transferred to the western coast across the western mountain ranges in India for generating hydropower between 1930–70; has become a matter of severe injustice both in the 1990s and in the 21st century when the per capita availability has fallen dramatically to less than 800 cubic meters per person per year. Similarly, sugarcane growers may establish appropriation rights in arid areas of a river basin, which appear to be patently unjust in times of severe drought.

(*d*) **Unequal Distribution of Social-Environmental Costs of Water Rights Development**. There are hardly any records of serious conflicts of free-flowing natural rivers. It is the upstream impoundment of water that leads to artificially created grievances between upstream and downstream users. This 20th century phenomenon is accompanied by unequal distribution of social and environmental costs of water resource development, leading to displacement, social disruption, ecosystem loses due to submergence of forest, impact on fisheries, changes in water availability in different water-regimes, unexpected floods due to shock-releases, etc.

Experiences in Conflict Resolution on the Bhima River Basin

The Bhima river is one of the main tributaries of the river Krishna, which forms a large river basin in the southern peninsula of the Indian subcontinent. It flows from west to east passing through the states of Maharashtra and Karnataka, before meeting the river Krishna, which flows further south to the state of Andhra Pradesh where it meets the Indian Ocean. The basin consists of four municipal corporations with a total population of 6,224,807 (*Census of India*, 2001). The main occupation is agriculture.

Establishing Multi-Stakeholder Platforms at the Local Level

Over the last 12 years, the Gomukh Trust has been engaged in carrying out watershed development activities in several micro-watersheds and sub-basins in the Bhima river basin. In addition, a large number of water users groups were formed on canals and lift-irrigation systems through the Panchayat Raj institutions and Gram Sabhas. Self-help groups amongst women were formed and the process of multi-stakeholder dialogue was initiated in the year 2000. Two large events were held in 2003 and 2005 where over 400 participants representing farmers (upstream and downstream), agro-industrialists and processing units, urban stakeholders, corporate sector representatives, government representatives from irrigation, water supply, pollution control board, revenue department participated actively. This formed the footsteps towards the change in societal perception from the local to Bhima river basin. Discussions and negotiations led to conceptual clarification, awareness about quantitative and qualitative dimensions, etc., an encouraging result was a consensus decision about not using the deep borewell for irrigation but only for domestic use. This was a decision based on awareness and negotiations, and did not require any compulsion or administrative enforcement.

Negotiating for Equitable Distribution between Upstream and Downstream Population

Since the promulgation of the Irrigation Act, 1927, the use of water from public dams and reservoirs in India was permitted only downstream of reservoirs by means of canals and distributaries for the purpose of irrigation and urban water supply. Till 1976, villages upstream of dams were not allowed to lift water for irrigation or for drinking purposes. In most cases, populations displaced by the dams and reservoirs were given cash compensations and left to their own resources/fate. These poor destitute dam oustees usually moved upwards along the hill slopes and eked out a living from the sub-marginal lands and decimated vegetation. On the other hand, the villagers downstream enjoyed the benefits of intensive irrigation. The lands and villages along the left and right bank canals emanating from the reservoirs turned green and prosperous. The assumption was that the water from reservoirs should be supplied to farmers who had been given commitments and not to others. In order to correct this anomaly, voluntary agencies and social workers initiated the process of lobbying with the irrigation department and the civil administration for a more equitable distribution of water.

Between 1978–85, a large number of meetings were held with the Pune district collector and chief executive officer for the purpose of negotiating for drinking water for villages upstream of dams to begin with, and then for crop protection. The dialogue process was long drawn out and persistent. The irrigation authorities first agreed to the lifting of 6 per cent of water from the reservoirs for utilization in the upstream villages. Drinking water supply schemes were worked out. Voluntary organizations like Gomukh Trust were working on minor irrigation schemes while the NGO 'Jeevan' was working relentlessly in the areas upstream of the Pawna dam from 1980–95 in areas upstream of several reservoirs like Khadakwasla, Panshet and Varasgaon dams, located about 45 km west of Pune city. The negotiations yielded incremental results.

By 1985, it was decided that the irrigation department would permit up to 6 per cent of the live storage in dams to be used by upstream farmers. And by 1995, the amount permitted for upstream utilization was increased to 12 per cent. Lift irrigation and farming on lands between the full reservoir (contour) level and the maximum draw-down level was allowed to some extent. The major share of such lifts was of course used by farmers who could lift the water to cultivate areas on higher contours — sometimes even up to 60 meter above the reservoir level. The advantages of these negotiations are now being reaped by people upstream as the government has announced a utilization of up to 20 per cent if the farmers use water-saving devices including sprinklers, drip-irrigation systems, etc.

Negotiating with the Forest Department

Background

Gomukh Trust is involved in comprehensive watershed development and management at several locations in Pune district; covering an area of 120,550 ha. of land. Although the Drought Prone Area Programme (DPAP) is being implemented in several states in the country, it is seen to be more effective in states, which have a more active involvement of competent voluntary organizations. The state of Maharashtra is one such example. Gomukh Trust, a planning and implementing agency for the DPAP has worked on 37,500 ha of land in over 200 villages, constructing check dams, farm ponds, wells and storage tanks, gabion embankments, contour trenches for ground water recharge, installing lift irrigation schemes and creating plantations with a diversity of indigenous plant species. The village water agencies were involved in the planning, designing, execution, monitoring as well as in the evaluation process during the period 1995–96 to 2002. Since January 2001, the village committees have taken up agricultural information technology extension in these areas. Since the core ideology for the development process in the Kolvan mini-basin has been the adoption of the participatory approach, involving the village community, 79 per cent of the expenditure (approximately equal to 1 million USD, calculated at purchasing power parity) was controlled and distributed by the village (Finance) committees by way of wages, about 9 per cent was spent on construction material and machinery and 12 per cent was for the administrative and technical support provided by the staff of Gomukh Trust.

Example I — Conflict between Forest Conservation and Development of Spring Tanks

One of the interventions — namely the development/augmentation of natural-spring-tanks required the permission of forest department as most of the springs were within the boundaries of protected (forest) areas like sanctuaries and national parks. Initially the forests department did not allow spring development because they argued that such development would divert forestlands for 'non-forest purposes' — which is patently in contravention of the Forest Conservation Act, 1980 of the Government of India. To overcome this situation, Gomukh arranged several meetings between the forest officials and the village elders to establish a dialogue for resolving the conflict. Gomukh advocated the view that the protection, augmentation and regeneration of the springs, achieved through soil and water conservation and plantation measures upstream would improve the ecological status of the forest by retaining soil moisture, increasing species diversity, providing much needed 'water holes' for the faunal species, increasing ground water level flora and improving sub-soil water percolation for recharging ground water aquifers. In addition, the spring water could provide an assured drinking and domestic water supply to downstream villages, thereby acquiring much needed support for the cause of forest-conservation. After a two-month long dialogue, and after agreeing to a number of preconditions and post-facto assurances the forest department finally accepted Gomukh's contention that the development of springs would not violate any laws, but would actually support the objectives of the Act in letter and spirit. The dialogue process achieved remarkable results. Not only did the forest department permit Gomukh to go ahead, but also the ministry of environment and forests asked Gomukh to prepare a full-fledged funding proposal for conserving the uppermost watershed of the Bhima Riverine Ecosystem. The work of augmentation of springs (six in number and covering about 800 ha. of land) have been completed and a full-fledged conservation plan has been submitted to the ministry for financial assistance equal to about Rs 100 million (USD 2 million measured in terms of PPP @ 2001 prices.) By 2004, the development of natural springs with tanks for water-regulation has been adopted as a special programme for the area covering the Western (Ghats) Mountain ranges by the government of Maharashtra.

Example II — Negotiating for Ensuring Minimum Flows and Rationalizing Urban and Agricultural Water-Use

With the ever-increasing population in the city of Pune and its suburbs, the demand for water in the metropolitan areas has been rising very rapidly. In the early sixties, about 200 million litres of water was abstracted each day from the Mutha river (via the right bank canal emanating from the Khadakwasla dam 16 kms west of Pune city). By the end of the century, the city was drawing about 600 million litres water each day. The same canal, which provides water to the city, is also responsible for irrigating about 60,000 ha. of land south-east of the city. Thus, an increase in the withdrawal of water for Pune city automatically has resulted in reduction in the supply of water for agriculture. In addition, about 350 million litres of sewage/liquid effluents is being released into the Mutha river each day, with a very small proportion being subjected to primary treatment. Thus, in addition to the conflicting demands for water between the urban and rural areas, the river also carries the sewage released by the city during the period October–June. This is so because there is practically no fresh water released from the dam upstream during the period.

To overcome these two serious issues, the city municipal council and the irrigation department entered into an agreement whereby 11 TMC of water will be supplied to Pune city by the Maharashtra Krishna Valley Development Corporation (MKVDC), and the Pune City Municipal Authorities would adequately treat the raw sewage so as to make it suitable for irrigation, and then pump the nutrient-rich water back into the irrigation canal for distribution to rural areas.

Unfortunately, the city failed to treat the raw sewage adequately and consequently was unable to pump it back into the irrigation canal, even though it continued to draw its full share of water from the river. When civil society organizations publicized these issues widely, the Municipal Corporation appointed an expert committee for preparing a Mutha River Improvement Plan (MRIP). The committee had the mandate to recommend solutions to solve this impasse. The author of this article, representing the Gomukh Trust was appointed on the committee along with other members. After a protracted dialogue lasting over 18 months, the city Commissioner agreed to install a series of six effluent treatment plants in 2001.

The dialogue process therefore yielded an important result, which would benefit the Mutha River Ecosystem and enable the continuation of irrigation in the farms downstream of the city. The dialogue continues and it is hoped that it would achieve two more important results, which are being negotiated at present. The civil society organizations are demanding the release of at least 50 cusecs of water from the dam for maintaining the minimum water flows and simultaneously improving the quality of water. In addition they are asking for the construction of a pumping station just near the eastern city limits for transferring a part of the treated water (about 300 million litres per day) to the irrigation canal. While the dialogue continues, a writ petition has also been filed in the Mumbai High Court for demanding these policy changes.

Creating a Basin-Level Platform

The activities of the Gomukh Trust during the past decade have been largely focused on surface water harnessing and management. Through participatory decision-making and watershed planning, it was realized that the success of the optimal utilization of water resources is possible only when ground water aquifers, the surface flows and the soil moisture are treated as one integrated entity.

In reality, however, the scientists and researchers working on ground water management (Ground Water Survey and Development Authority of the Government of India and its state-level agencies) have often been at loggerheads with the state irrigation department, each trying to claim its superiority and importance in being able to solve the problems of water scarcity. The citizens at large have been playing the role of silent onlookers while the two state agencies continued to quibble.

Realizing the futility of such quarrels, the Gomukh Trust along with other voluntary agencies decided to initiate and establish a periodic dialogue between the agencies and all other stakeholders. Gomukh also realized that taking up issues and resolving conflicts was important and that it was equally important to broaden the dialogue process and establish a forum for discussing matters of policy which had not yet reached the status of a conflict. To initiate the dialogue, the issue chosen for discussion was the National Water Policy, the importance of 'dialogue' and the alternative approaches to water management. All concerned stakeholders and representatives were invited for the first round of the dialogue in the second week of February 2002 in Pune. As organizers, we felt encouraged that all state agencies dealing with water and a large number of civil society representatives attended this dialogue and discussed several issues without inhibitions.

Being encouraged by the intensity and richness of the first dialogue in February 2002, Gomukh announced a second round in August 2002 where we discussed the Draft Maharashtra Water Policy released in the previous month, and the drafts of two proposed acts, namely the Maharashtra Farmers Participatory Irrigation Management Act, 2002 (MFPIM Act) and the Draft Maharashtra Water Resources Planning and Regulation Authority, 2002. This two-day workshop also attracted an equally good response and it was widely reported by the press and the media. The event kicked off public discussions regarding 'privatization of water' v/s 'social ownership of water', equitable distribution of water, citizens' participation in decision making, pricing of water and so on.

The upshot of the initiative was that even on intangible issues of policy, apparently conflicting parties and even citizens in general are willing to join a dialogue if the process is sufficiently participatory, and if it continues to acquire social credibility through the media and press publicity. What we often heard from the participants was that they had all wanted such a dialogue but were waiting for others to initiate it. An important lesson we learnt was 'somebody has to bell the cat'.

Upscaling to River Basin Partnerships

The experiences mentioned above led to a situation by the end of 2001 where it became necessary to establish a multi-stakeholder partnership in the Upper Bhima Basin. At the same time, the India Water Partnership under the auspices of Global Water Partnership was trying to identify a basin suitable for establishing an Area Water Partnership. The idea was mooted by some of the retired-bureaucrats of the state irrigation department, and it was immediately supported by civil society organizations, representatives of urban and industrial interests, ecologists, advocates representing people, displaced by dams, and many others.

This is probably the first time that a dialogue has been initiated between practically all stakeholders interested in either developing or using the water of the Bhima system, who have come together and formed a water partnership. A vision document for the period 2001–25 was negotiated through a dialogue and was presented at the South Asia Water Forum held at Kathmandu in the last week of February 2002, and later on in Vietnam during a 'Dialogue on Water for Food and Water for Nature'.

The Bhima Partnership was followed by a multi-stakeholder conference organized by Gomukh, where well over 400 participants representing several stakeholder interests attended the deliberations. This led to media coverage, public discussions and general people's awareness about transboundary basin issues.

Prerequisites for an Effective Participatory Process for Resolving Conflicts Related to Sharing of Transboundary Rivers

1. The parties concerned (states or stakeholders) should feel the need to negotiate or alternatively civil society mediation and awareness through information dissemination should bring the parties to a peer-level platform to articulate the need for conflict resolution.
2. The negotiation has to start with a base document, which clarifies all facts and scientific information and frames the contentious issues for discussion, clarification and negotiation. The base document must contain a basin-level water balance study.
3. A knowledge base should constitute both the prevailing laws and policies, nature of demand for water, variety of users, their quantities, accepted possible priorities, hydrogeological status and ecological regimes on the one hand; and social perception, historical antecedents,

idiomatic and anecdotal and literary perception on the other. In addition it should have statistics and quantitative analysis offering possible implication of existing water uses, current and best practices; inefficiencies, wastages, technology for improvements and optimizations of water-use, instruments and mechanism for equitable water distribution, etc.
4. The establishment of the structure of forum for discussion and meeting opportunities for creating awareness on micro and macro issues, and procedurally established rounds of well-designed and iterative discussions among all conceivable stakeholders. This will help to build capacities for informal/formal negotiations, and a culture of dialogue rather than confrontation and parochial jingoism.
5. Creating a tradition of declaring and publicizing such discussions and negotiations. This includes periodical discussions and negotiations at the village level, district and regional level, sub-river basin level, and river basin level (i.e., vertical progression and upscaling.)
6. Similarly carrying out negotiations as confidence building measures among all stakeholders for deepening and broadening the understanding of river basin issues and cross-sectoral linkages.
7. Emergence of a facilitator, initiator who acquires credibility and acceptability among the stakeholder as a neutral arbiter who can maintain an open, clear, transparent and inclusive process. The facilitating agency/CSO (negotiation ambassador!) has to ensure that the process is non-directive and unbiased on voluntary collaboration.
8. Build bridges of communication and understanding across and within sectoral perspective.
9. Raise financial support (fund) from the contending stakeholders to increase ownership of and accountability to the process and outcomes of the dialogue. This will ensure autonomy of the procedures.
10. Create peer-level moral and societal pressure so that the contending parties agree to abide by the decisions rather than rejecting and disregarding them.

Conclusion

The problem of inter-state transboundary sharing of water is not about inadequacies in legal and policy internments or physical parameters, but about the near absence of a *'participatory-democratic-culture'*, capable of achieving negotiated agreements. Once such process for achieving this prerequisite takes place through the most elegant and simple principle, namely *'equitable allocation of water'*, it will be sufficient for determining fair shares of water between riparian states on any river basin.

References

Committee on the Uses of the Waters of International Rivers. 1967. *The Helsinki Rules on the Uses of the Waters of International Rivers.* London: International Law Association.

Government of India. 1956. *Inter-State River Water Disputes Act.* New Delhi.

Hirsch, Danielle and Vijay Paranjpye. 2005. *River Basin Management: A Negotiated Approach.* Amsterdam and Pune: Both ENDS and Gomukh.

Joy, A. K. J., Bhisham Gujja, Suhas Paranjpe, Vinod Goud and Shruti Vispute. 2007. *Water Conflicts in India — A Million Revolts in the Making.* New Delhi: Routledge.

Ministry of Water Resources. 1999. *The National Commission on Integrated Water Resources Development Plan.* New Delhi: Government of India.

Office of Registrar General and Census Commissioner. 2001. *Census of India.* New Delhi: Ministry of Home Affairs, Government of India.

United Nations. 2005. *The Millennium Ecosystem Assessment.*

9

Negotiation Through Social Dialogue: Insights from the Cauvery Dispute

S. Janakarajan

The Context

The Cauvery is one of the most important rivers of peninsular India. Karnataka and Tamil Nadu are the major states staking claim on the Cauvery water, while Kerala and Pondicherry are the other riparian states, which benefit in a small way. Therefore, Cauvery is an inter-state river as per the provisions of the Constitution of India. Cauvery river basin is spread over an area of 87,900 sq.km which accounts for nearly 2.7 per cent of total geographical area of the country. The basin covers an area of 48,730 sq.km in Tamil Nadu and 36,240 sq.km in Karnataka. The river travels through a distance of 800 km before reaching the Bay of Bengal in the southern Tamil Nadu coast.

Tamil Nadu has had a much earlier and more rapid history of development of irrigation command than Karnataka. In the case of Tamil Nadu, the initial pre-Mettur command area of 14.4 lakh acres was added with 11.4 acres during the successive plan periods. Therefore, the total command area of 25.8 lakh acres in Tamil Nadu state was developed over a long period of time. Whereas, the irrigation development in the Karnataka part of the basin was only 4.42 lakh acres as on 1971 and it was only after 1971 that schemes were drawn to bring new areas under irrigation. The underlying point is that the millions of Tamil Nadu farmers and landless agricultural labourers have had access to Cauvery water for centuries. Any denial of the hitherto access to this precious resource may result in serious cultural setback in the region.

Appreciation of the doctrine of 'prior appropriation', an international water law is quite central in this context. According to this law, the first user who puts the water to beneficial use acquires a prior right to the extent of such use. On the contrary, if the upper riparian state is allowed to expand irrigation development quite late, without considering the irrigation network already developed by a lower riparian state, not only it creates tension, but subsequently it might also lead to a situation in which a lower riparian state would hardly get any access to water. That is why it is vital to protect the 'historical interests' of a lower riparian state. But at the same time, we cannot stop the upstream state from developing their irrigation command. This is the crux of the dispute.

Cauvery river has got the distinction of being the most disputed and litigious river in contemporary India. Whenever monsoons fail, the conflict between the two major riparian states explodes; at times even taking violent turns. Indeed, the inter-state dispute has prompted the Supreme Court and the Government of India to seriously consider the possibility of linking of Himalayan rivers with that of the rivers in peninsular India (what is popularly known as Ganga–Cauvery link).

The Cauvery water dispute between the riparian states is quite different from other inter-state water disputes such as Krishna, Godavari or Narmada. In the case of the latter, the disputes revolve around the utilization of the untapped potential, whereas, in the river Cauvery, the dispute is still around the issue of sharing the available water in the river, although the available potential has already been utilized to the maximum extent possible by all contending states. As per the present utilization pattern the total water demand in the basin is about two times the potential available. This is indeed the key to the dispute. Two issues emerge from this argument: one, the Cauvery water dispute therefore, should not and cannot be compared with other inter-state water disputes, and two, whatever potential that may have already been developed should be protected whether it is in the upper or lower riparian states. In this sense, the Cauvery water dispute is around the issue of not sharing but re-sharing of the river water (Guhan 1993).

The events (of the agricultural season 2002–2003) that followed the Supreme Court's directives to the Karnataka Government to release at least 0.8 TMC ft of water to the Tamil Nadu state has almost landed us in a constitutional crisis. This is unprecedented in the history of any inter-state water disputes in India. Thanks to the strong words of the Supreme Court, the chief minister of Karnataka not only tendered an unconditional apology for having disregarded the Supreme Court's directives, but has also started releasing water. There was a sigh of relief from many quarters, in particular from the concerned civil society of both the states, not because the released water would contribute to saving of crops of Cauvery farmers in Tamil Nadu, but simply because a major constitutional crisis was averted.[1]

Distress conditions had often led to violence in the past, the worst form of which was witnessed in December 1991, when thousands of Tamils and their properties were targets of attack in Karnataka. Whenever tensions erupt between both states, more than Cauvery farmers, the urban middle class and anti-social elements in rural and urban areas take advantage.

A prudent analysis of the long-standing and a widely debated dispute between Karnataka and Tamil Nadu exposes the sense of distrust that they have towards each other. Karnataka, despite its delayed utilization of the Cauvery water, believes that its legitimate entitlement to use water is questioned; Tamil Nadu, on the other hand, has had a much earlier and a more rapid history of development of irrigation command in the Cauvery basin and as a lower riparian state feels that it is at the receiving end — both literally and metaphorically, this anxiety is a direct result of the state having to bear the brunt of the burden of floods, drought and pollution. *But none of these burdens negates the fact that both the states' food production depends so massively on the availability of water in the KRS and Mettur dams, and that the livelihoods of millions of people in the basin area are contingent on the flow of water in the river.* While Tamil Nadu finds it devastating and thorny to be at the mercy and goodwill of the Karnataka government during times of scarcity, Karnataka feels that in the absence of water for its farmers it is difficult to release water to the downstream state. The Tamil Nadu government has sought the intervention of the Supreme Court for a clear title of its share to the Cauvery water.

The establishment of the Cauvery Water tribunal in 1990 as per the provisions of the Constitution of India (Inter-State Water Disputes Act, 1956) was very eagerly awaited to give a permanent solution. The tribunal gave its interim award according to which Karnataka was supposed to release a minimum of 205 TMC ft water every year. Karnataka rejected the interim award. After over 15 years, the tribunal also announced the final award in February 2007, which was also rejected by Karnataka. Although Tamil Nadu articulated no major objections to the award, subsequently it expressed a few reservations. Seemingly, a total expression of victory would mean that the party of the other is defeated. Therefore, a statement of satisfaction may not be forthcoming from any state. What is however worrying is the petitions filed by all the states in the Supreme Court. The Supreme Court has also admitted the petitions, which means that the struggle would go on for some years. Ideally, in a federal structure such as India, the tribunal's award (whether interim or final) shall be both decisive and binding for all parties concerned; otherwise, it would result in unpleasantness, and eventually may even result in a constitutional crisis. Nevertheless, given the past history and hard positions already taken by both the states, it is unlikely that the final award would end the dispute. This is not to be construed as a tone of cynicism but only as a note of insurance.

Some Perspectives on Social Dialogue

What one encounters in India is the overuse of both surface and ground water, pollution of surface and ground water bodies, growing scarcity and competing demands for water across different sectors, lack of safe drinking water, growing conflicts and competition among various users/uses, and most of all, mounting bitter inter-state water disputes. The government(s) try to address all these issues by and large through what one may call 'the conventional wisdom', namely, supply augmentation measures and judicial interventions. Indeed, the Government of India's move towards linking of major rivers in the country has been the latest in this direction.

As a matter of fact, judicial activism, remnants of which could be witnessed in a few historic judgments delivered by the highest judicial authorities of India, could not also travel too far due to various compulsions.[2] Under these *vulnerable circumstances*, and *when all efforts seem unyielding*, what one needs is a platform for a sustained dialogue for all those concerned. This is what is referred to as *multi-stakeholders' dialogue (MSD)*, a widely advocated measure all over the world for resolving volatile and explosive situations in natural resource management. Furthermore, in the wider context of water, food and environment, the dialogue initiative presents an alternative thinking in water management. The alternate thinking aims to initiate a dialogue process, to draw together all stakeholders in a selected river basin to a common platform. The dialogue process ultimately aims to bring about changes in the overall welfare of people through amicable settlement of disputes and through the sustainable use of water in all sectors such as agriculture, industry, domestic and in the protection of ecology.

After all, why is one interested in the multi-stakeholders approach to achieve 'sustainable use and development'? There are policies and legislations to protect and regulate the use of water. Some feel that these policies and legislations are less effective; other stakeholders feel that they favour some while prohibitively curbing the interests of others; and some others think that the policies and legislations are exclusively designed to please some constituent elements of the users. This kind of a vulnerable situation breeds competition, conflicts, extreme events, and eventually leads to unsustainable development.

The MSD approach helps not only to find solutions to conflicts in a given point of time but also provides a platform for a continued dialogue for sustainable use and development. It provides an opportunity for all users to get together not only to examine the nature of problem confronted by users and by society at large, but also to understand and appreciate each other's problems. In addition, it provides an opening and mindset for the multi-stakeholders' group to understand each other's strengths and weaknesses, their interdependent status in the society and their overall contribution to the country and to the economy. Most important of all, *the MSD offers a cordial and congenial climate where conflicts could be turned into opportunities* for an effective and fruitful collaboration for sustainable development and for the general well-being of the society. Overall, in the MSD process, one can grasp the attributes of the various stakeholders, generate insights by studying the nature of competition and conflicts prevailing among them, document strengths and weaknesses and the degree of stakes enjoyed by them, and above all it can help in the construction of supply and demand scenarios for future use with some degree of accuracy, authenticity and endorsement.

Social dialogue approach would not yield instant results. It is a process in which all stakeholders, though initially fight and debate, settle down only after a while. What is however, crucial is to sustain the tempo and concern showed by stakeholders until some tangible outcome emerges. *Nonetheless, it needs to be emphasized that the intention of MSD is more to complement rather than to replace existing measures of dispute resolution and water management. Recognition and support of state machineries and legal mechanisms are necessary conditions for the success of any MSD process.* Yet, to set in motion the MSD is not a simple and straightforward issue. Congenial atmosphere must be created for stakeholders to come together and to hammer around all key issues. Any concerned individual with adequate knowledge on the subject could play the role of a catalyst.

Disputes in Sharing of Natural Resources

Scarcity is the most fundamental reason for the occurrence of disputes. In the particular case of water, scarcity grooms not only intense competition and conflict, but also poses a grim threat to the entire developmental process of an economy. In a most general sense, disputes may be defined as 'irreconcilable interests' of individuals or groups or society at large, which are expressed in terms of hostile attitudes, claiming of exclusive rights or pursuing their own interests through actions that jeopardize the well-being of other parties. Such disputes in India are more delicate and threatening. Indeed, many disputes are metaphorical in nature, which take added dimensions when seen in relation with multitudes of maladies such as poverty, hunger, malnutrition, unemployment, migration, lack of or inadequate health care, lack of protected drinking water supply, high illiteracy and demographic pressure in addition to caste differentials. All these are manifested in unsustainable use and bitter disputes among users of natural resources.

Primary and Secondary Disputes

In disputes such as those, which relate to natural resources, it is vital in the first instance to differentiate between primary dispute and secondary dispute. For instance, in the case of the Cauvery water dispute, while the primary dispute is inter-state in nature (primarily between Tamil Nadu and Karnataka), the secondary dispute is intra-state in nature (between head and tail enders and across different sectors such as agriculture, drinking water users, industries, etc). In which case, it is important first to arrive at a solution for the primary dispute before attempting on resolving intra-state disputes. Such an approach is imperative for, "one has to know the size of the cake before dividing it among various legitimate claimants within the state". The other way would be to take up all disputes — both primary and secondary — simultaneously for a dialogue process. The second approach may not produce desired results since there may be a tendency to mix-up issues during the dialogue process. Therefore, in tricky and explosive situations such as the Cauvery dispute, utmost need is necessary to build confidence among the farming community as a whole (from of all riparian states), who are the most affected lot and to create an enabling atmosphere for a dialogue process to takeoff. Once the 'primary dispute' is 'under control', one can set in motion the dialogue process for resolving 'secondary disputes'.

The Main Contending Issue and Rationale for Social Dialogue

What is the rationale for promoting a dialogue between farmers of Tamil Nadu and Karnataka? As indicated earlier, farmers have been listening to what the political parties and bureaucrats have been saying in this matter. Thanks to the efforts of the politicians the communication gap between farmers of Tamil Nadu and Karnataka has increased! The information flow is in fact severely restricted. On the contrary, wrong data and misinformation have further widened the gap between the actual water users of the respective states. Following are some of the issues on which there exists communication gap either due to a lack of information flow or due to a lack of understanding.

First, farmers in Karnataka believe that the Thanjavur farmers grow three crops in a year using the Cauvery water; by three crops, they refer to *kuruvai, thaladi* and *samba*. But these are not crops but seasons. While the first two are short duration seasons, the last one is a long-duration season. Given the available water (which never exceeds 8 months' supply even in the best years), Tamil Nadu farmers can grow either two short-duration crops (*kuruvai* and *thaladi*) or one long duration crop (*samba*). As a matter of fact, the crop months of all three seasons work out to 14 (4+4+6). Therefore, it is impossible to grow three crops in all three seasons using the Cauvery water. This is an ill-conceived notion, which prevails primarily due to a communication gap between the farmers of both states. The dialogue process would help in bridging such gaps.

Second, there is a feeling among farmers of Karnataka that Tamil Nadu has massive unutilized ground water potential in the delta region and that they are unwilling to shift their crop pattern from paddy to less water-intensive crops. Moreover, it is often stated that Tamil Nadu farmers need to adjust their farming to the growing conditions of scarcity. First of all, the data published by the Central Ground water Board gives exactly the opposite indication. As of 1992, groundwater development in Nagappattinam district is 100 per cent (eastern part of the Cauvery delta). Blockwise information in this district indicates that in six-out-of-eight blocks ground water is over-exploited, while in the remaining two blocks categorized as 'dark blocks' the ground water development is in the range of 85–100 per cent. It is obvious that in the last 10 years the situation would have worsened. Second, until 1931 when the KRS dam was commissioned, the utilization of water from the Cauvery river for agriculture was entirely dependent upon the river course or run-off from the river or in the form of water diverted through the anicuts. In other words, water kept flowing downwards and flooded the delta region of the lower riparian state. The long history of flooding and water logging has resulted in soil salinity and as a consequence the land in this delta region has become unsuitable for any crop other than paddy. Annual crops like sugarcane, banana and oil seeds are cultivated in the western part of the delta using canal water supplemented by ground water, where water logging is not an acute problem (viz., Thanjavur district).

Third, people in Karnataka believe that a good deal of water is wasted in the Tamil Nadu part of the delta. This is absolutely baseless because there are many structures specifically designed and constructed in the Tamil Nadu delta with a view to re-using the drained water which otherwise would go waste into the sea.

Fourth, there is an absolute necessity to educate farmers of both the states about the relevance and significance of our federal structure.

Fifth, there is a need to promote a dialogue with a view to differentiating 'use rights' from 'exclusive rights'. The problem arises mainly because of the feeling of such exclusiveness in particular among farmers in Karnataka. They were misguided for long and needed to be educated. It is the responsibility of the governments in both the states to educate farmers that no single state has an exclusive right over the Cauvery waters. Until now, all the political parties have failed in their duty to educate farmers about this distinction. Each successive government in both the states has spent its energy in politicizing the issue, thereby promoting regional chauvinism. Therefore, a continuing dialogue among the farming community seems to be a useful way out for the current stalemate.

Sixth, the dialogue is absolutely necessary to undo all the hitherto built-up apprehensions and misgivings and to create a sense of caring and sharing as well as a climate of warmth to promote the feeling of brotherhood.

There may be more reasons why dialogue between farmers of both states should take place. But the need for a 'non-official initiative' such as the one suggested in this proposal is only the first step towards finding a lasting solution. Such a dialogue could then provide a favourable climate and mindset for the farmers of both states to accept the final award of the tribunal.

Why and How Cauvery Water Dispute is Different from Other Inter-State Water Disputes in India?

First of all, Cauvery river is a heavily used river basin not only for agricultural purposes but also by other competing uses such as drinking (both rural and urban) and industries. Most neglected use has been ecology and environment. Therefore, there is no unused water potential in this basin, which is indeed the main source of contention in the case of other inter-state water disputes. Thus, the dispute in Cauvery is around the issue of re-sharing of heavily used river water. Second, since the Cauvery water dispute has already got a long history, it has become an intense political issue by becoming enmeshed in electoral politics. Many elections have been fought primarily on this issue. No political party could afford to ignore it and for that reason were often compelled to champion the cause of Cauvery farmers in both states. Finally, the river Cauvery like Ganges, is deeply embedded in cultural and religious practices of the people of both states. Many religious festivals in the basin areas of both states centre around the river. Any move therefore to stop the river flow hurts the people's sentiments and their emotional attachments to the waters of the Cauvery.

How are the Main Interests of Contending Parties Reflected in their Involvement in the Dialogue Process?

The main contending states are Karnataka and Tamil Nadu, and in a small way, the state of Kerala and the union territory of Puducherry. The vexed nature of the dispute has prompted farmers of both the main contending states to participate in the dialogue. Farmers of all contending states fully realize that key water management measures featuring modernization of canal network, construction of modern control structures, etc., are long pending due to the ongoing dispute. The best way to resolve an inter-state water dispute in a federal set-up within the framework of democratic governance is to seek legal recourse. Tribunal is the natural outcome under such circumstances. Although tribunals were instrumental in resolving several inter-state water disputes in India, the Cauvery dispute stands tough and remains hard-hitting. The tribunal's interim as well as the final award has not contributed to the resolution of the dispute so far. If at all, it has created more tension, especially in Karnataka. Not only farmers of the two contending states, but also the media and the civil society at large look forward to a fruitful dialogue and resolution of the dispute in a peaceful manner. The resolution passed in the very first meeting held on 4–5 April 2003 at Chennai summarizes their involvement.

We agree to care for each other, share each other's problems and also agree not to indulge in counter productive activities. We also endorse this initiative, affirm our faith in dialogue and commit ourselves to its progress.

Objectives of the Cauvery Multi-Stakeholder Dialogue

Our vision in proposing a multi-stakeholder dialogue in the Cauvery is as follows:

(a) To bring together farmers of the riparian states on a common platform for a fruitful dialogue.
(b) To take a pragmatic view of the current situation in the Cauvery basin.
(c) To reduce differences and the communication gap among the riparian states.
(d) To set in motion healthy information flow among the riparian states.
(e) To undo the apprehensions and misgivings built-up over time and to create a climate of warmth, sense of caring and sharing and to promote the feeling of fraternity.
(f) Most of all, to find a way forward for the benefit of the *entire Cauvery family* in the larger interest of the country.

The main participants of this MSD process are to be farmers drawn from the principal riparian states. In addition, some academics, retired bureaucrats, NGOs, lawyers, people from the media, and other concerned citizens were also invited to participate. As a beginning for the dialogue process, two meetings of farmers' representatives of both the states have been held, one in Chennai and the other at Bangalore. The Chennai meeting took place on 4 and 5 April 2003 and was attended by 110 participants while the Bangalore meeting took place in the same year, on 4 and 5 June, which was attended by 120 participants.

How does One Proceed? What are the Steps to be Followed to Arrive at a Solution?

Some spade work was necessary before organizing the dialogue between farmers of both states—the key step was hydro-historical research and documentation. The need of the hour is to understand the dispute in a socio-economic, historical and cultural perspective. Subsequently, it was felt necessary to tour the delta areas in both states in order to meet, discuss and promote the idea of the dialogue with key leaders of farmers' organizations, individuals, such as retired judges, NGOs and some political veterans. While research is still ongoing, touring in the basin areas took six full months. This involved visits, individual and group meetings and documentation. The third step was to hold two major meetings of farmers' (and a few others), one in Chennai (held on 4–5 April 2003) and the other in Bangalore (4–5 June 2003). The *Committee of the Cauvery Family* was formed immediately at the end of the second meeting with 32 members from both states. This Committee has met nine times so far in different parts of the basin in both states and have also visited different segments of the Cauvery basin area in order to gain some first-hand information for the first time.

What Solution(s) Could be Arrived at?

No solution has been reached so far, but we have been quite successful in defusing the tension. Furthermore, we are currently developing scenarios of water use by using the WEAP (Water Evaluation and Planning System, a hydrology model developed at the Stockholm Environment Institute, Boston). We are also in the process of producing a film on Cauvery to disseminate the message of negotiation through social dialogue to the grass-roots in both the states.

Achievements Worth Mentioning

(a) Brought together varying interest groups on a common platform that was found to be a tough job.

(b) Created greater understanding of the problem among farmers of both states.
(c) Created an opportunity for farmers to meet on a common platform to share their problems; this has developed the feeling of brotherhood and camaraderie among farmers and helped to undo the hitherto built-up prejudices and hatred; redefine issues in the larger socio-economic and cultural perspectives and above all facilitated a process to resolve the dispute by adopting a more scientific approach.
(d) The media reported after the ninth meeting of the Cauvery Family held at Tiruchi on 9 April 2006 as follows: *The mutual visits of farmers' leaders to the command area in the Cauvery basin, it is learnt, have removed suspicion over the irrigation practices, which have often caused heartburn among the farmers...the mutual visits have helped in clearing the air about the misconception and lack of faith in each other* (*The Hindu*, 13 April 2006).

What are the Lessons Learnt that Would be Useful for Others?

- Sound research is a necessary condition for undertaking and carrying the dialogue forward.
- Degree of success or failure of dialogue initiatives depends upon active and sustained state support.
- A threshold level of crisis will make dialogue initiative more sustainable and will ensure active participation of all contending stakeholders; otherwise, only one set of stakeholders will participate.
- Need for an untiring facilitator who can carry on with the job of facilitating and arranging a platform for the dialogue to continue.
- Dialogues are never smooth; there will be lots of ups and downs; this should be expected.
- Final outcome is uncertain; difficult to judge—but in the absence of a viable alternative there is a case for pushing the dialogue initiative as far as possible until one reaches any where near a viable solution.

- Any decision arrived at by means of farmers' dialogue could be put into practice only through due political process. Therefore, it is necessary that non-governmental/non-political initiatives of this kind get the recognition of political parties and government. Furthermore, in a federal set-up any discussion or decision concerning inter-state water dispute or river basin planning should reflect the views and concerns of multi-stakeholders' dialogue.

The Role of Facilitator

- Facilitator should be untiring, committed, socially responsible and be ready to face ups and downs.
- Should be agreeable to all members of the committee.
- Should always be open-minded, neutral, impartial and have the capacity to carry forward the dialogue process under all circumstances.
- Should be qualified to mediate on any said issues; he/she should have credibility and be trustworthy.
- Should be able to design the initial phase of MSP and MSD.
- Should be able to mobilize resources when needed.
- Should always be alert while conducting meetings so that all bitter and irrelevant arguments can be avoided.
- Should be able to contribute a great deal to the documentation process.
- Should ensure that meetings are held in a cordial atmosphere.
- Should facilitate meetings in respective areas of members of the committee and should also facilitate field visits in all contending areas.
- Should ensure that members take active interest in arranging committee meetings and field visits.
- Should always drive home the fact that the committee's mandate is to arrive at a consensus and find ways forward and not argue, and accuse each other or the government.
- Should always promote the feeling and welfare of the committee — as a family — rather than as individuals.
- Should never interfere in internal matters of members and their associations; but at the same time should make sure

that such differences do not interfere in proceedings of the committee.
- Should be in a position to delineate key and important issues from unwanted and irrelevant issues; should promote discussion on all relevant issues if it is time consuming.
- The best strategy for the facilitator is to capture the common points of agreement and build on them; should postpone all issues on which agreements are deficient.
- Should never impose or force any viewpoint on members even if they are extremely relevant and important.
- And, finally, the facilitator should project himself/herself to be a positive person and never ever give any grounds for negative thinking.

Some Open-Ended Questions

- How to sustain a spirited dialogue process among multi-stakeholders until one reaches a tangible solution?
- What is the time frame involved for achieving a reasonable degree of success with tangible and innovative solutions?
- Should one wait until one reaches 'the disaster/threshold level' to initiate a dialogue process?
- Do we have the necessary enabling environment to initiate MSD process in the context of the Cauvery impasse?

Some Post-Final Award Scenario

All contending states have approached the Supreme Court by way of filing what is called Special Leave Petitions. The tribunal itself was constituted as per the provisions of the Inter-State Water Disputes Act 1956, as amended in 2002, after considerable hearings in the Supreme Court. Therefore, the Final Award of the tribunal is considered as good as the verdict of the Supreme Court. If that is so, it seems to be unclear as to why the Supreme Court has again entertained the Special Leave Petitions by the contending states instead of referring them back to the tribunal. Now that the dispute is back in the Court, it may take some more years before the final verdict is announced. Nevertheless, our concern is to what extent the *so-called* final award is going to be accepted by contending states, in particular Karnataka. The legal course has already taken one full round without the sighting of any prospect. The ongoing social dialogue gains more significance precisely under these circumstances.

Notes

1. However, the situation that followed the releasing of water was quite grim in Karnataka: farmers' violent protests had caused enormous damage to the public property and the state of affairs had compelled the Karnataka Government to impose curfew in Mandya. The Congress MP from Mandya had even sent his resignation letter.
2. For instance, the Supreme Court judgment in the tannery pollution case in the Palar river basin in Tamil Nadu was historic and quite strong. But despite such a strong judgment pronounced in the year 1996, untreated effluent from tanneries continue to be dumped in the Palar river: The Judges pronounced, 'It is no doubt correct that the leather industry in India has become a major foreign exchange earner and at present Tamil Nadu is the leading exporter of finished leather accounting for approximately 80 per cent of the country's export. Though the leather industry is of vital importance to the country as it generates foreign exchange and employment avenues, it has no right to destroy the ecology, degrade the environment and pose as a health-hazard. It cannot be permitted to expand or even to continue with the present production unless it tackles by itself the problem of pollution created by the said industry', AIR 1996, Supreme Court 2715, paragraph: 9.

Reference

Guhan, S. 1993. *The Cauvery River Dispute: Towards Conciliation.* Chennai: Frontline Publications.

10

Integrated Water Resource Management (IWRM): An Alternative Paradigm

Anitha Kurup

Introduction

At the global level, the universal right to water across species, and more importantly across generations, was asserted by the 1977 UN Water Conference declaration that everyone has 'the right to have access to drinking water in quantities and of a quality equal to their basic needs'. More recently, the UN, in 2002, declared that water is 'a public good fundamental for life and health. The human right to water is indispensable for leading a life in human dignity. It is a prerequisite for the realization of other human rights'.

The increased demand for water globally, due to the rising population coupled with the rapid pace of development has resulted in an increased attention on water as a resource. There are more than 263 watersheds that cross the political boundaries of two or more countries. These international basins cover 45.3 per cent of the land surface of the earth, affect about 40 per cent of the world's population and account for approximately 60 per cent of the global river flows (Sandra 1999).

Over the past decade, India has witnessed perhaps the largest vagaries of monsoon coupled with an increased demand on water with large-scale expansion of not only cultivation but also the simultaneous growth of industries and an increased population. This has further been made more complex by the impact of globalization, where there has been a widespread introduction of new crops along with new technologies, often making comparatively greater demand on water.

There has also been a widespread realization that in the event of meeting the food requirements of the country, the green revolution has also left large areas of land uncultivable with increased salination resulting in the degradation of land. Given the above concurrent developments, it has almost become imperative that India not only manages its river water but moves to a much broader base of developing an integrated management of its land and water resources.

Significance of IWRM

The last decade in particular has given an increased emphasis on the understanding of the management of water resources using the IWRM paradigm. This paradigm has gained increased currency due to its adaptation and endorsement by powerful international agencies like the Global Water Partnership (GWP) and the World Water Partnership (WWP). If one were to follow the evolution of this terminology over the years, it is more than evident that the meanings assigned to it has been expanding to include aspects that hitherto did not find a place within its restricted definition.

The technical advisory committee of GWP has adopted the following definition; IWRM is a process which promotes the coordinated development and management of water land and related resources, in order to maximize the resultant economic and social welfare in an equitable manner without compromising the sustainability of vital ecosystems.

There has been a paradigm shift in water resource management that has been led by an increased realization that land resource management is intrinsically linked to water resource management and the sustainability of water resources has a dynamic relationship to land use and it's users. The definition of the users has also undergone large-scale changes to not only include the present but also the future generations. The ambit of the users even within the category of the present generations has increased to encompass within the category of farmers, marginal farmers and landless agricultural labourers. The primary users of water, is no longer restricted to the farming community and now includes domestic users, industries and the whole of humankind also extending to plants and animals.

Changes in land use not only effect water resources but may have long-term effects on land resources, and the subsequent environmental degradation may have a direct impact on the poor.

The non-availability of clean drinking water coupled with low nutritional levels, makes the poor, and particularly women, the worst targets.

Thus, the emergence of users with multiple identities and as belonging to different systems — national, state, political, social, economic class, and caste has also impacted the meaning of IWRM. Hence, the broad definition of IWRM will have to particularly draw attention to the following facts:

1. Water as a resource does not confine itself to surface water alone but ground water, and even more importantly, rain water. There is the need to view water as having an integral relationship with land and not in isolation. Furthermore, land and water are parts of a larger ecological system and hence the management of these resources needs to be seen from that vantage point.
2. Water as a resource is a part of multiple systems. These systems are political, technical, social, economic, cultural, rural/urban, gender-based and include user groups, all of which operate at international, national, state and local levels. Synchronization of the different systems in the IWRM paradigm will have to be the underlying principle of IWRM.

Water Related Disputes — Selected Examples

International River Basins

In formulating a broad definition of IWRM, it may also be useful to examine selected river basin disputes, both at the international as well as regional/national level. A critical analysis of this can lead one to look at the range of conflict resolution mechanisms within an illustrative mode rather than embark on an exhaustive analysis.

◆ *Nile River Basin*

The Nile river basin flows through nine countries, which covers one-tenth of Africa. The basin is characterized by different political identities, different climatic conditions and represents

Map 10.1: Nile River Basin

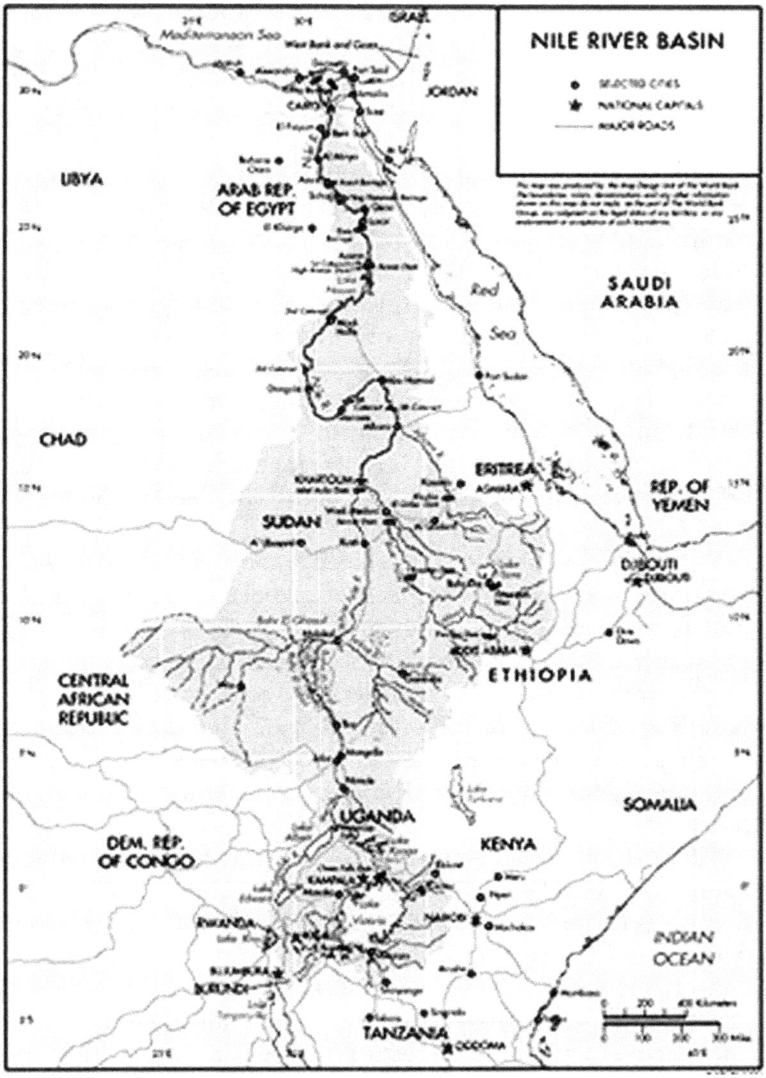

Source: http://siteresources.worldbank.org/EXTAFRNILEBASINI/Images/map_sm_2.gif (accessed 18 February 2009).
(© The World Bank. Used with permission.)

different social and religious groups that together form the community of the Nile basin. The conflict of the Nile basin can be traced to the historical exploitation of one riparian state, namely Egypt, over the other eight countries, characterized by the traditional, single-minded engineering approach towards river and water management.

The recognition of traces of conflict in the early twentieth century was when Sudan increased its use of the upstream water for irrigation purposes. Since the contenders for the use of the Nile were Egypt and Sudan, an agreement for sharing of water was drawn between them. However, when many more riparian states staked claim to the water resources of the Nile in tune with the needs of their own growing populations, Egypt, who had traditionally enjoyed the exclusive rights started experiencing a disadvantage. While the demand for water gradually increased there was a realization that the development of the river basin will be a relatively better proposition than sharing the already available water.

Among other things, treaties, basin organizations, commissions and other innovative initiatives have all become a part of conflict management in the Nile basin. Inter-basin organizations like the Kagera, and later, the Nile initiative in 1992 were launched to promote cooperation and development of the basin. The Nile River Basin Action Plan in 1994 defined 22 development projects covering broad themes like water-resource planning and management, regional cooperation and environmental protection.

While the potential of the Nile river is largely untapped in most countries covered by the basin, the member countries affected by the issues related to equitable sharing are five of the ten poorest countries of the world. Thus, in the long run, the interface of conservation and management of water resources, technology for efficient use, improved sustainable agricultural practices and efficient use of water for domestic purposes, equitable distribution and ecologically sound practices for sustainable water use for the present and future generations will require better planning at the basin level. What stands out in the river basin management in relation to the Nile basin, is the sustained interaction of the countries forming the basin and the evolution of mechanisms towards the development and management of the basin.

◆ *Mekong River Basin*

The first dimension of the Mekong basin is to place its development within a historical context. The fundamental principle involved treating the river basin as two parts; the lower Mekong basin consisting of Cambodia, Laos, Thailand and Vietnam and the upper basin that facilitates the separation of China. The exclusion of China and the process of development of the lower Mekong basin along the lines of donors and the UN that had a predominant presence can lead one to believe that river basin management in this case could well have been used as a geopolitical tool.

The recent thrust of the development of the Mekong basin has acquired a visible shift from a focus on its hydraulic and hydropower capacity to that of sustainable development. The language used by donors and countries that are spearheading this change emphasise values like democracy, social justice and equitable distribution which are in direct contradiction with the culture of the political elites of the riparian countries.

The river basin organization is expected to be instrumental in devising and implementing a sound policy. The policy is expected to cover ecological, hydrological, social and political aspects. The legitimacy of this body is derived by the ability it has to negotiate with the governments of the riparian countries. In the case of the Mekong basin, the riparian states often view that the river basin institutions clash with the sovereignty of the member states. The divergent views or for that matter the vision of the development of the Mekong basin arise from the asymmetry in the production and access to environmental information.

Indian River Basins

Indian rivers can be broadly classified into two kinds, the first being the Himalayan rivers and the other being the rivers of peninsular India. The Himalayan rivers include the Ganges, Indus, Brahmaputra and their tributaries and are created by the snowmelt and glaciers of the Himalayan ranges along with the catchments of

Map 10.2: Mekong River Basin

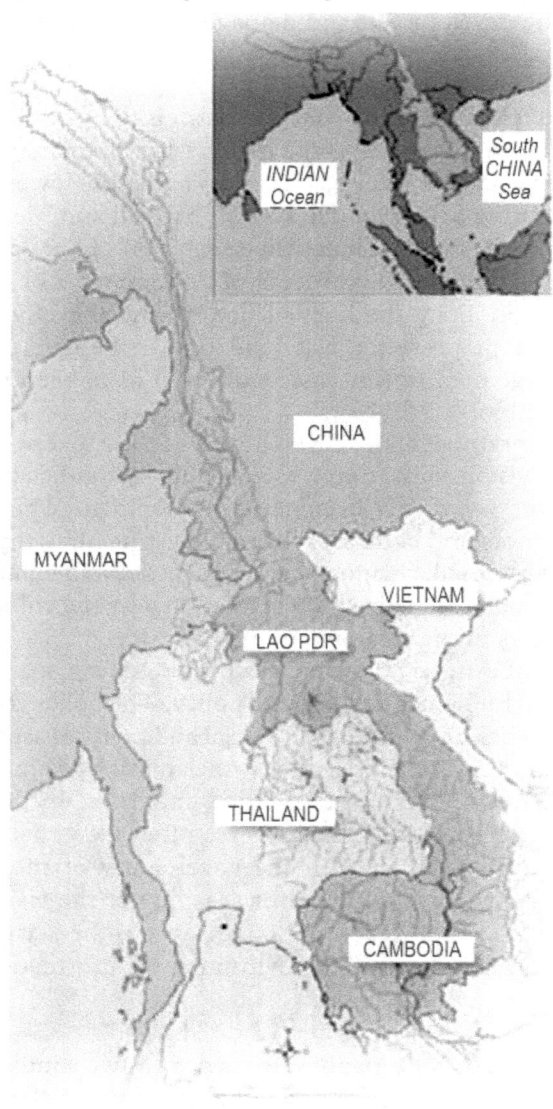

Source: http://www.mrcmekong.org (accessed 18 February 2009).
(© The Mekong River Commission. Used with permission.)

rains in these regions. On the other hand, the rivers of peninsular India comprise Krishna, Godavari, Narmada, Mahanadi, and Cauvery and their tributaries. These rivers that flow from the Western or Eastern Ghats depend on the rains that fall in the catchments areas.

The Himalayan rivers are characterized by being not only inter-state rivers, but also rivers that flow between countries. Thus, the sharing of water is not only within the states of the Indian territory but also between the neighbouring countries. For instance, the Indus river water is shared between India and Pakistan; the waters of Brahmaputra between India and China, and the Ganges between India and Bangladesh.

The fact of the matter is that while, there have been conflicts between the different states associated with the Himalayan rivers, the process of resolution of these conflict has always had the inherent advantage of the waters being shared at the international level between two or more countries. Resolution of water conflicts have in most cases not been an issue of conflict amongst the different nations, since there were relatively larger conflicts these nations had to deal with, and hence, water conflicts have taken a backseat. Given this historical fact, the guiding principles of international water sharing have to some extent had an impact on the principles of water sharing between the states within India.

Unlike the Himalayan rivers, the different states through which water flows in peninsular India have seen conflicts between the upstream and downstream states. Historically, with uneven development between the different states; the downstream states are relatively more developed and have used the water available to their advantage. The upstream states have most often than not engaged in development through the construction of dams and reservoirs at a later point in time that has resulted in the fear among the downstream farmers, that water that was available to them in the past will be affected with the new constructions that will divert the flow of water. Table 10.1 gives selected examples of national water-related disputes.

The resolution of the Krishna–Godavari conflict was primarily aided, since at that point of time the three states had Congress government. But more importantly, the conflict resolution was attempted at a point of time where the demand for water by the

Table 10.1: Selected Examples of National Water-Related Disputes

Names of the River Basins	Names of States Sharing the Water	At the Time of Conflict Resolution	Political Situation during Conflict Resolution	Reference Point of the Tribunal Judgement
Krishna and Godavari	Andhra Pradesh, Karnataka, Maharashtra	Sharing of water that was under-utilized at that point in time. Absence of state chauvinism.	All the basin states were under the Congress government and the central government was also under Congress rule.	The Tribunal Award was based on a more recent agreement (after 1951). The dispute between the two states were over irrigation, hydroelectric generation.
Cauvery	Karnataka, Tamil Nadu, Pondicherry and Kerala	Resharing of waters that are already being fully utilized.	The conflicting basin states have governments who are of different parties but also belong to different alliances at the centre. The period also saw the exit of the single largest party rule at the centre. Emergence of increased complexities of centre–state relationships.	The Tribunal Award was based on an agreement (1924), where reorganization of the states did not take place. The dispute is largely for water for irrigation.

(Table 10.1 Continued)

(Table 10.1 Continued)

Names of the River Basins	Names of States Sharing the Water	At the Time of Conflict Resolution	Political Situation during Conflict Resolution	Reference Point of the Tribunal Judgement
Ravi–Beas	Punjab, Haryana, Rajasthan	Part of the larger river of Indus that has the history of the Indus treaty. Water conflict on use of water for irrigation by all the states. Sharing of water that was under-utilized at that point in time. Allocation based on the current use and allocation based on the projected use of water.	The process of re-looking at distribution of water was initiated by the chief ministers of Haryana, Punjab and Rajasthan in 1981. Central government referred the dispute to a tribunal in 1986. The period also saw the exit of single largest party rule at the centre. Emergence of increased complexities of centre–state relationship.	The Tribunal Award is based on agreements as recent as 1981. The dispute about the inclusion of Haryana and Rajasthan as part of the river basin did not stand.

Source: Richards and Singh (2001) and Wolf et al. (2005).

Map 10.3: Indus River Basin

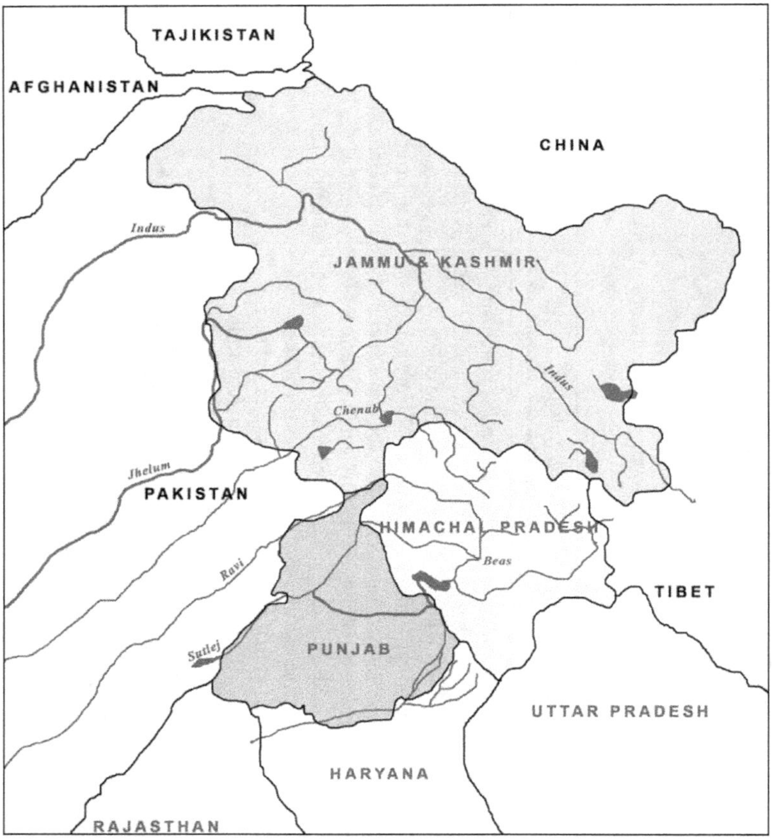

Source: Adapted from Central Water Commission of India: http://cwc.nic.in/regional/chandigarh/images/basin.jpg (accessed 18 February 2009).*

three states was well within the limits of the available water. It may also be noted that the resolution of the conflict of this river basin was primarily drawn by continued negotiations among themselves and reached an agreement on all disputed issues. This agreement was of a more recent origin.

Unlike the Krishna–Godavari conflict resolution, the setting of the tribunal, in the case of the Cauvery dispute, was at a point of time that witnessed conflicting governments in the two states, as well as the relationship each of the states had with the central

Map 10.4: Cauvery River Basin

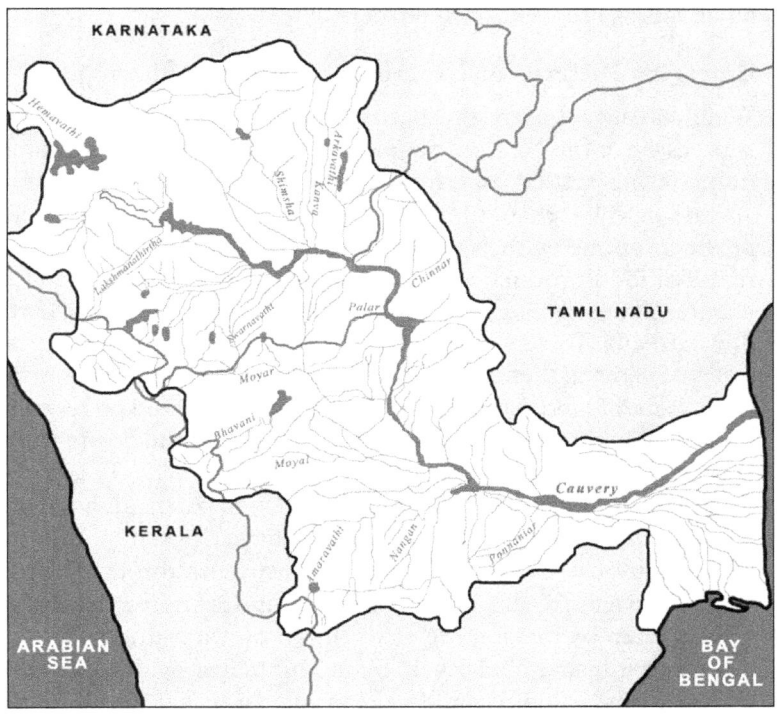

Source: Adapted from Dr Eberhard Weber, School of Geography, University of South Pacific, Fiji. http://www.usp.ac.fj/fileadmin/files/schools/ssed/geo/staff/weber/cauvery4c.jpg (accessed 18 February 2009).*

government. It must also be noted, that the time at which the tribunal was constituted, the Cauvery river basin was already overused. Hence, a change, or for that matter a decision on the distribution of the Cauvery water would have affected either of the states since, the demand projected by the states for the Cauvery water was well beyond the water that was available in the basin.

The principle of equity also ran into severe problems in this case — simply because one was caught between the existing practice of cultivating a third crop in the downstream states and the demand of the upstream state for the development of cultivation that will allow the farmers to take a second crop. Also, the historical disadvantage to the state of Karnataka by its identification with the

Madras Presidency was also cited for the unfair treaty that gave an advantage to the Tamil farmers.

Emerging Patterns of Experiences of Water Sharing

Water has always been considered as one of the non-negotiable resources since the Neolithic age when human beings seized to be nomads and settled on bank of rivers. It was during this time, around BC 8000 that cultivation was seen as an avenue for food production and with it, grew the rising need for water in the realm of food production.

Internationally, it can be observed that countries located at the downstream of rivers often complained of the upstream riparian states of denying them access to water especially in the event of development of land for cultivation. The process could be reversed if the development of the land for cultivation of the downstream riparian states followed that of the upstream states. The fact of the matter is that development in general, and particularly in agriculture and other related activities that have a potential of using water resources like urbanization, industrialization and hydro-electric power projects, has been uneven and is affected by several factors, thereby making the distribution of water unequal.

At the international level, Syria complains against Turkey, Egypt against Ethiopia and Pakistan against India. Within India, one has also witnessed similar situations, where Tamil Nadu complains about Karnataka regarding the Cauvery waters and Karnataka does the same against Andhra Pradesh in the case of the Krishna waters.

While the examples above trace the existing mechanisms used in the resolution of water sharing in river basins both in the international and regional/national context, they also point out to the roles that several contenders play in actively negotiating their right for water through their different identities. Conspicuous though is the absence of women, who also constitute a vast constituency of water users, but may be the crucial link that will make the realization of IWRM a reality. Also, the role of traditional water management systems that focus on water conservation, has not been explored. This may be critical in resolving the persistent water conflicts within the country.

Complexities of Water Resource Management

Water as a resource is intrinsically linked to other resources like land, air, forest, plants animals and human beings, and cannot be viewed and hence managed in isolation. There has always been a considerable debate on arriving at a viable unit to examine these inter-linkages, such that, it provides one with the access to look at mechanisms of management and distribution. For a long time, national and state boundaries were considered as viable units for water management that has been largely responsible for lending the political overtures to the discussion of water distribution even to date.

History has established beyond any doubt, that this unit of analysis comes with its own sets of problems. Simultaneously, around the same time, there has been a lot of effort to visualize the world as a global village, and the fact that interdependency between countries is not only growing in the area of economic trade and development but also in areas like education, health, agriculture and industry. The twenty-first century has also laid siege on issues related to water, environment, climate change where there are increased pressures for cooperation and conflict resolution between nations.

Also, within a river basin every thing is connected. Surface water, groundwater, the quality and quantity of water are all central to the distribution and management of water. There is also widespread fluctuation in water availability, both with regard to space and time. This is further complicated which in its current form is fragmented and subjected to vague demands both at the level of quantity and quality. At the general level, despite differences about geographical area, the variations in the economic development of the different countries, the nature of the conflict remain by and large the same.

Socio-political Dimension of Water Resource Management

Historically, water resource conflicts have always fundamentally been constructed with an engineering approach. In an approach of this kind, construction of dams and distribution of water in terms of quantities has assumed central attention. It has also been noticed that conflicts between the states in India very often has been on issues such as geographical boundaries or river water sharing. Both of these are fundamentally traced to land and water resources and the identity of the state is reiterated through the regional division based on linguistic criteria, which has been the predominant principle in dividing the states in Independent India.

It has been argued that equity and social justice be the guiding principles for the distribution of water between the competing states. However, the need to adopt the equity principle to govern the internal distribution of water within each state has often been seen as something that shifts attention from the primary problem. It may be useful at this point, to question the very construction of the problem as a binary between primary and secondary. It also raises the fundamental question of whether the secondary follows the primary.

The complexity of the problem is further enhanced with increased compartmentalization of the issue as one, that deals with distribution versus management and further, through the isolation of technology use that can aid more efficient management of water resources. For more reasons than one, one is led to believe that the water conflicts between the states in India, and in this particular case, the Cauvery river conflict, reiterates regional identities to give it a more alive and emotive status.

Further, given the reality that the construction of the problem is not only between two states that have had a conflict historically where language became the bone of contention, but also that during this point of time, it also represents state governments with different political alliances at the national level. The decision about the sharing of water between the two states can be also constructed as benefiting one state, while being contrary to the interests of the other state. However, whatever the nature of the decision, the underlying implication is that there is no resolution to the conflict to be seen in the near future.

Alternatively, if one were to focus attention on making sure that state identities may not in itself be a necessary tool, for the reason that the construction of the Karnataka farmers/Tamil farmers is illusionary and the sub-categories within these two broad groups will include small, medium and landless labourers on the one hand, and farmers belonging to various castes/class groups as well as female and male farmers on the other.

The process of redefining multiple identities also bring to the fore the need to recognize users other than farmers, who also have a right to the water resource. This may include domestic users both for drinking water and sanitation, as well as industries and urban households that also become a part of the stakeholders who claim the right to use the water. In these redefined groups, it may be likely that there will be more effective participation of the new stakeholders, who may include issues of integrated land and water resource management where technology will become an integral part of conflict resolution.

By increasing the base of water users, it also leads us to look at a more fundamental problem of managing, rather than distribution of a resource. These exercises will in whatever way, if implemented, at least meet the needs of a greater proportion of stakeholders if not completely meet the requirements of the user base that is fast expanding.

The pace of development over the last decade in the developing countries has given rise to new challenges particularly in the management and distribution of resources. Water and land have been identified as critical resources and have been receiving increased attention. Developing countries like India are actively being encouraged to move from the traditional supply-side orientation to demand management under the broad framework of IWRM.

One must also take into cognizance the vested interests of different stakeholders to keep alive the issue of conflict to further their ends. Thus, one can visualize the construction of the problem of water sharing. So, very often focus on how many cusecs need to be released is determined by the increasing demands of the farmers of the concerned states. This problem in its current form can be stated as:

1. The emphasis is on the distribution of the available water with an increasing area of land coming under irrigation especially with changing cropping patterns that inherently require more water.

2. The distribution principle has often constructed two contenders vis-à-vis the farmers of the two states — all the while arguing that they represent the interests of the farmers from each of their respective states.

Water Resource Management — Shifting the Focus

Redistribution of resources, both natural, human and material is increasingly been seen as essential for the overall development and progress of humankind. Amongst the several resources, water has been identified as an important resource that not only requires urgent redistribution both internationally and at the national level, but calls for a much more informed definition of a resource that is fast depleting. The setting of the millennium development goals (MDGs) and its dynamic relation to water distribution at the global level has been receiving increasing attention over the recent past. The MDG framework within which one locates the distribution of water resources is useful since it enables one to include the most deprived sections of society.

The management of a fundamentally important resource like water has gained significance in the wake of the realization that this resource is fast depleting, and it is time one focused attention on not only the distribution of water resources among the living humankind but also extend it for the future generations.

Castro (2005), in his article on water, power and citizenship, has brought to the fore yet another dimension that is important to better understand the changing nature of water conflicts worldwide. His fundamental argument looks at the overarching tendency of mainstream explanations of water crisis that tend to reduce the problem to its economic, technological and physical-natural dimensions thereby undermining the socio-economic and political inequalities that are central to water conflicts across the globe. This has been the case since most water debates have had a domination of engineers and technocrats, who with their respective disciplinary training are able to view the problem in its technical dimensions. Therefore, the dominant paradigm that exists currently in understanding the management of water resources is technological in its character. The lack of independent studies with a social science perspective has resulted in the slow development of conceptual and theoretical tools that can capture the interplay of factors that are both natural and social.

Another dimension that needs considerable attention is that water scarcity is not a supply-side problem only, and that it is very often socially constructed. The nature of the conflicts that arise out of water scarcity that is also socially constructed assumes a dynamic nature that makes it inherently difficult to define the problem as a static one. This occurs because the very problem of the construction of a conflict that can have several dimensions of complexities. Any conflict resolution strategy must as its first step be able to arrive at a preliminary definition of what this conflict is, what issues they encompass, and more importantly what values they reveal. The conflicts are different in nature, scale and in terms of what players they involve. They also demonstrate a dimension of space and time.

Women and Their Experiences

The prevalent approach to water resource management stems from the fact that it has always been the domain of technocrats and engineers. Given this as the underlying premise, there has obviously been greater emphasis on quantitative and so-called scientific approach to understanding the problem of water. This not only removes or isolates the problem from the societal context within which it is located, but is also operated through a process that is complex and well beyond the purview of existing scientific tools and techniques.

Unlike natural phenomena, social phenomena has been a relatively new subject of study. It is also beyond any form of dispute, if one were to say that some of the basic principles that guided the study of natural phenomena like universality, replicability, cause and effect are not so simply delineated when one studies social phenomena. The definition of the subject matter under study as well as the role of the knower or investigator is significantly different. This, in a way, will not only give rise to a body of knowledge that includes the experiences of subjects who represent the constituents of the use of this vital resource namely water but may perhaps lead to an alternative paradigm that may be useful to resolve the conflict.

The participation of women in the management of land and water resources, very often has overlooked the divergent needs and interests of not only rural women but also of the urban poor women. The need to treat women not as a single category also deserves equal attention. This is also true of poor men who constitute the under-privileged sections of the society. The need to construct the participation of women beyond numbers and their cursory/quiet representation on committees to a more proactive participation where their needs take centrality vis-a-vis issues of water management is urgently required.

To further illustrate the point, the evolution of the water users association as a mechanism to increase the canvas of participation is a laudable one. In that, the one-third representation given to women in some of these bodies also needs appreciation. Nevertheless, one would like to bring into focus the following:

1. Are these new kind of organizations, that in a sense constitute the WUAs characterized by inequality, what are these forms or does one see the replication of the power equations of the larger society within these organizations.
2. How effective are the representation of women on these bodies, and what are the opportunities given to women to actually steer, define and make a visible shift regarding the activity of these informal bodies like the WUAs.
3. At the most fundamental level, is this body of water users' association organically linked to the decision-making process of water sharing either at the sub-basin level; river basin level or for that matter sharing of water at the state, national and international level.
4. In particular, what are the negotiating powers that these water users' associations have, to take decisions regarding water sharing beyond the existing paradigms of reference? In fact, do the water users' associations question the efficiency of the existing paradigm of water sharing? If they do, are they vested with power to redefine the existing paradigm used for water sharing to include those sections like themselves, underprivileged, poor and women in a much more substantial sense?

Traditional Knowledge Systems

Women should be factored into water resource management not only for reasons for equity but also because of the fact that they are critical players in traditional knowledge systems. Most of these experiences have not been documented systematically and hence remain oral histories. In most subsistence societies, women were an integral part of agricultural systems, and the division of domestic needs and agricultural needs were not as stark as it is today. The absence of women's experiences from the formal documentation process that did not provide a space for oral histories resulted in the systematic elimination of a large part of traditional practices that focused on the roles of women and other indigenous groups that were excluded from formal knowledge systems.

While it can be argued that during the past, water conflicts have not come to the fore since the demand for water was manageable with limited population growth. This cannot perhaps preclude the traditional water management systems that may have been in place despite the absence of demand. One may also have to trace historically, if the concept of water management systems had an inbuilt bias of also representing the so-called developed countries, the urban middle class and other socio-economic groups that have been the forerunners of both the construction of the problem and the consequent management of water resources as they see it. In this, there is a strong possibility of losing the critical experiences of those traditional, perhaps smaller groups that were excluded either consciously or unconsciously. In such a context it may also be useful to critically examine the practices of water management where women played a crucial role.

Understanding local knowledge systems is fundamental to discern the complexities of water resource management of any region. It also becomes equally important to trace historically the intricacies of gendering this knowledge through the interplay of power which then becomes the basis through which the process of construction, valuation and distribution of knowledge takes place within the realm of water resource management. The engineering-based approach to water resource management that dominated the field till very recently also tended to marginalize women's experience and knowledge. A much more equitable interaction of the traditional knowledge base with the technical know-how may pave way to new directions of water management systems. In the wake of increased challenges not only to the water management systems in particular but also in relation to development theories, there is an increased focus on socio-cultural contexts. The traditional knowledge systems are perhaps a rich resource that were rooted in and emerged in response to the socio-cultural realities. By its design of being local, the participation of women and other groups hitherto not a part of formal knowledge systems found space in this system. The alienation of the traditional knowledge system is thus reflected in the present scientific knowledge that attempts to provide an overall understanding of the water management systems that dominate the literature on water. It has been argued that the existing scientific knowledge base whether with respect to water resource management or otherwise is abstract, invariant through change in space and location and is transmitted formally, while the traditional knowledge system is holistic, contextual-adaptive and transmitted informally (Blaikie et al. 1977).

However, Agarwal (1995) argues that both systems of knowledge have their individual strength and are equally dynamic and contextual, socio-culturally bound and value laden with specific histories and particular burdens from the past.

First, that traditional systems are extremely important even in this so-called world of modern science and technology. Their ecological rationality remains valid even in the modern context. Among other areas in which India's traditional knowledge systems have developed and survived from pre-historic to contemporary times is that of the development and management of water resources. Management of water resources historically has not been restricted to river basin but has also extended to lakes and tanks. A close look at the range of human activities, including agriculture, animal husbandry, different types and levels of economic activities including the cooperatives reveal very clear patterns of the availability of water and its distribution within specific regions.

Towards an Alternative Paradigm

The challenge in the present-day context of managing and distributing water has been recognized and hence the integrated water resource management paradigm has emerged. The overemphasis on technological solutions and the underplay of the social dimensions particularly in the implementation phase has led to the kind of problems we face today. The mode of response to the gender and social dimension of the questions has always been peripheral and less visible at the field level. This paradigm, while enlisting the principles of equity and social justice does not lend itself sufficiently to accommodate the socio-cultural specificities of different regions of the world. This is particularly true in the case of regions that have had a long history of the managing water before the emergence of the IWRM framework. Their contribution to the development of this paradigm has not been systematic or for that matter all encompassing so as to reflect their rights and interests because of the inherent dominant framework that has both a western and a technological bias. The inherent absence of knowledge systems of a relatively large population including that of women who have no access to the formal knowledge system makes the very construction of the problem within this paradigm limiting. Thus, attempts to resolve the issues are half-hearted, and do not address the real problems of a large section of the society who are excluded from this whole process.

Seventeenth century onwards till date has seen a clear shift from traditional water resource management to more technology-based water resource management resulting in the construction of large dams and transfer of water from river basins through canals to the most controversial mode of interlinking of rivers as one of the solutions to the problems of water resource management in the country.

Alongside, although on a smaller scale, there has also been the revival of interest in traditional water systems in recent years, both for theoretical and practical purposes, especially by development activists (including organizations like the Centre for Science and Environment (CSE), Alwar's Tarun Bharat Sangh (TBS), etc., and people like Anna Hazare), scientists, environmentalists and many others associated with the cause of sustainable development. The increased role of community and community-based knowledge has challenged the present understanding of water management systems. Consequently, there has been increased interest in several parts of the world to tap on the traditional knowledge practices and evolve ways to integrate it. These systems of managing the environment constitute an integral part of the cultural identity and social integrity of many indigenous communities in our country. The knowledge gained by these communities is through conceptualizing empirical observations to better understand nature and thus interpret and predict it (Iaccarino 2003).

Unlike Western science with its positivist leanings, that favour analytical and reductionists methods, the traditional knowledge systems attempt to study reality not as a linear conception of cause and effect, but rather as a world made up of constantly forming multi-dimensional cycles, in which, all elements are part of an entangled and complex web (Freeman 1992). Comparing two complex systems of knowledge like western science and traditional knowledge with particular reference to water resource management, has inherent problems.

It may also be useful at this point to objectively analyse the strengths and limitations of both western science and traditional knowledge systems that are characterized by their individual approaches, systems of construction of knowledge, methods used, validation procedures and hence attempts at a symbiotic relationship between the two in a way that leads to mutual respect and willingness to dialogue on the problem of water resource management. Thus, the process must explore the possibility of including multiple viewpoints that are vicarious in building a cognitive universe and can disclose a complex picture of reality (Mazzocchi 2006).

While it is true that western science is developing new approaches to studying complex systems, it will be equally useful to learn how traditional approaches explain such complexities. Fundamentally, western science and traditional knowledge constitute different worldviews of knowledge but are rooted in the same reality and hence can mutually supplement our understanding of the world. However, this process of supplementing has to take care to ensure the absence of the inherent hierarchy between these two kinds of knowledge. The process should reflect mutual respect and learning and make sure there is an emergence of an approach that derives its strength by the preceding process of mutual respect and learning from the traditional knowledge systems and the western sciences that take care of the current complexities and competing demands. This, in turn, can be utilized to meet the present-day needs of rural and urban areas.

Western science uses analytical and reductionist approaches, whereas local knowledge systems use a wholistic perspective with a predominant systemic linkage approach that is also intuitive. Both these approaches have their strengths and weaknesses. However, developing an approach that combines these principles so as to make sure that one derives the maximum benefit for most people, poses a challenge that needs to be addressed. This will perhaps result in an efficient use of resources so as to cope with increasing population pressures, emerging needs of newly constituted groups, scarcity, fluctuations in the environment, or other contingencies that are part of this new developing world. Achieving sustainability in resource management may require a continuous dialogue between the western modern approaches alongside with a holistic traditional indigenous system that may have stood the test of time. Space for nuanced understanding by smaller independent initiatives that are flexible, sensitive and respond to realities at the field level needs to be created while one moves towards sustainable resource management.

Conclusion

The socio-political character of water resource management that draws on the principle of equity and justice and broad basing the stakeholders to legitimately include groups that represent the user community, in particular women and the poor, may provide an analytical tool to address the persisting water conflicts within India. Knowledge of indigeneous systems of water management may also have a useful contribution to make to the understanding of the construction of water crisis that is more often than not stated within its technical framework. Thus, water conflicts must be viewed as a part of a social process where inequalities of power in terms of class, caste and gender identities characterize the social fabric of the community who constitute the primary stakeholders.

Note

*Acknowledgement with thanks to Mr Raja P. K. for the adapted version of the pictures of the Indus and Cauvery river basins.

References

Affeltranger, Bastien. 2005. 'Intra-basin Conflict Resolution in the Mekong Basin: Is a Reconciliation of Water Values Possible?', *Extended Abstract* at the International Workshop on Value of Water — Different Approaches in Transboundary Water Management, 10–11 March, Koblenz, Germany.

Agarwal, A. 1995. 'Dismantling the Divide Between Indigeneous and Scientific Knowledge', *Development and Change*, 26(3): 413–39.

Blaikie, P., K. Brown, M. Stocking, L. Tang, P. Dixon and P. Sillitoe. 1997. 'Knowledge in Action: Local Knowledge as a Development Resource and Barriers to its Incorporation in Natural Research and Development', *Agricultural Systems*, 55(2): 217–37.

Calder, Ian R. 2004. *Blue Revolution: Integrated Land and Water Resource Management*. Earthscan Publications.

Castro, Jose Esteban. 2005. *Water, Power and Citizenship: Social Struggle in the Basin of Mexico*. UK: Macmillan Publications.

Freeman, M.M.R. 1992. 'The Nature and Utility of Traditional Ecological Knowledge', *Northern Perspectives*, 20(1): 9–12.

Iaccarino, M. 2003. 'Science and Culture: Western Science Could Learn a Thing or Two from the Way Science is Done in Other Cultures', *EMBO Rep* 4: 220–23.

Iyer, Ramaswamy R. 2004. 'Beyond Drainage Basin and IWRM: Towards a Transformation of Thinking on Water', 6 March–26 April. Center for Global, International and Regional Studies, University of California, Santa Cruz.

Jacobs, Jeffrey W. 1995. 'Mekong Committee History and Lessons for River Basin Development', *The Geographical Journal*, 161(2): 135–48.

Kelkar, Meghna. 2007. 'Local Knowledge and Natural Resource Management', *Indian Journal of Gender Studies*, 14(2): 295–306.

Mazzocchi, Fulvio. 2006. 'Western Science and Traditional Knowledge', *EMBO Reports* (5).

Postel, Sandra. 1999. *Pillar of Sand*. New York: W.W. Norton & Company.

Richards, Alan and Nirvikar Singh. 2001. *Inter State Water Disputes in India: Institutions and Policies*. Research Paper, University of California.

Wolf, A. T., A. Kramer, A. Carius and G. D. Dabelko. 2005. 'Managing Water Conflict and Cooperation', in *State of the World 2005: Redefining Global Security*. Washington, DC: Worldwatch Institute.

About the Editors

N. Shantha Mohan has a Ph.D. in Education and is presently Professor, School of Social Sciences, National Institute of Advanced Studies, Bangalore. Apart from having taught in the postgraduate Department of Education, she has been active as a researcher and activist on issues relating to gender and governance, with special reference to social justice and human rights. She was the consultant in charge of mainstreaming gender issues in the Indo-Dutch Tungabhadra Irrigation Pilot Project, Phase II from 1995–97. She also facilitated the establishment of the Zonal Water Partnerships across the country to address concerns relating to the water sector. Currently, Professor Shantha Mohan is engaged in developing a toolbox on IWRM and Transboundary Water Conflict Resolution, as well as actively influencing policy changes on issues relating to transboundary water sharing and application of the Convention for the Elimination of All Forms of Discrimination against Women (CEDAW).

Sailen Routray is a doctoral student of Development Studies in the School of Social Sciences at the National Institute of Advanced Studies (NIAS), Bangalore. Before joining the doctoral programme at NIAS in August 2005, he worked with the Tata Institute of Social Sciences (TISS), Mumbai for a year, first as a research assistant and then as a research associate, from 2004–05. He holds a Master's degree in Social Work from TISS, Mumbai. Apart from writing for various academic journals, he has co-translated Arundhati Roy's *Greater Common Good* into Oriya. His Oriya translation of Ivan Illich's *Deschooling Society* is under production.

N. Sashikumar has a Bachelor's degree in Civil Engineering and a Masters degree in Remote Sensing. He is currently working for his doctoral degree at the Indian Institute of Science, Bangalore.

He has been engaged in a number of projects involving remote sensing applications to water resources. Besides creating models of urban water supply systems, Mr Shashikumar also is involved professionally with a number of civil society organizations such as those focusing on literacy for educating children, and farmer's organizations for water management.

Notes on Contributors

Ramaswamy R. Iyer is Honorary Research Professor at the Centre for Policy Research, New Delhi. A former civil servant, he was Secretary, Ministry of Water Resources in the Government of India (1985–87). He has been a member of the Sardar Sarovar and Tehri Projects (1993–97) and the National Commission on Integrated Water Resources Development Plan of the Government of India (1997–99) and is the Chairman of a Task Force on Natural Resources, Environment, Land, Water and Agriculture, set up by the Commission on Centre–State Relations. He has been a consultant for the World Bank; the World Commission on Dams, IWMI, UNDP, the European Commission and others. Currently (since August 2007), Professor Iyer is a Member of the UNSGAB High-Level Expert Panel on Water and Disaster, an adjunct to the UN Secretary-General's Advisory Board on Water and Sanitation. He has published extensively on water-related issues. His books include *Water: Perspectives, Issues, Concerns* (2003) and *Towards Water Wisdom: Limits, Justice, Harmony* (2007). He is currently editing a book on *Water and the Laws in India* to be published shortly.

S. Janakarajan an economist is currently a Professor at the Madras Institute of Development Studies (MIDS), Chennai. He has a Ph.D. from MIDS, University of Madras, Chennai and a Post-Doctoral degree from Cornell University, and subsequently worked as a Visiting Professor at the Queen Elizabeth House, Oxford University for a year. His areas of interest are development studies, rural development and agrarian institutions, climate change and disaster risk reduction, water management and irrigation institutions, conflicts and conflict resolution, environment, urban and peri-urban issues, and markets. Professor Janakarajan is the author of the *Cauvery Family*, (an initiative towards conflict resolution) which he started in 2003 that has brought together farmers of Karnataka and Tamil Nadu to resolve the Cauvery inter-state water dispute.

R. Jeyaseelan is former Chairman of the Central Water Commission. He has a Master's degree in Technology from the Indian Institute of Technology, Madras and a post-graduate diploma in hydropower development from the Norwegian Technical Institute, Trondheim, Norway. He has wide experience of about 40 years in various facets of water resources development and management. His field of specialization relates to hydropower development and has handled several prestigious projects in India and abroad. He has travelled widely and handled consultancy assignments in Afghanistan, Laos, Burma, Mozambique, Sultanate of Oman and Indonesia, some of them funded by the Asian Development Bank and the World Bank. He has been a member of several government delegations and was Vice President of the International Commission on Irrigation and Drainage. He has contributed over 30 technical papers and delivered a number of lectures. At present he is engaged as a Consultant in the water sector dealing with hydropower projects, in particular.

Anitha Kurup is Associate Professor, School of Social Sciences, National Institute of Advanced Studies, Bangalore. An educationist by training, she has engaged in extensive field-based research and theoretical analysis on issues related to education and gender. Her current research interests are in the area of evolving alternative conceptual and theoretical frameworks in relation to interdisciplinary research questions spanning the fields of water, education, gender and philosophy. During her course of work, she has also developed new methodologies that helps engage with complex but real life problems. Her thesis on interaction of village characteristics and school was published as *Village, Caste and Education* in the year 2000. She also co-authored the book *Status of Rural Women in Karnataka* in 1998.

Narendar Pani is Professor and Dean, School of Social Sciences, National Institute of Advanced Studies, Bangalore. An economist by training, he has worked in both academic and media institutions, including Indian Institute of Management, Bangalore and *The Economic Times*. He has, over the last 30 years, written extensively on a variety of subjects. He is the author of *Inclusive*

Economics: Gandhian Method and Contemporary Policy (2001); *Redefining Conservatism: An Essay on the Bias of India's Economic Reform* (1994); and *Reforms to Pre-empt Change: Land Legislation in Karnataka* (1983). Professor Pani is also the author of a primer on the WTO, two novels and a booklet on theatre, besides being a regular contributor to newspaper editorials.

Vijay Paranjpye holds a Master's Degree in Economics and Political Science, and an M.Phil in Development Economics. In 1971, he started his teaching career at the Ness Wadia College, Pune and later taught Environmental Science at the School of Environmental Studies, University of Pune from 1992–06. His main research interest has been in the area of economic and environmental evaluation and assessment of large infrastructure projects like dams, metropolitan water supply systems, mines, thermal power stations and airports. Prof. Paranjpye's areas of interest include management of national parks and sanctuaries; alternative approaches to water resource management with special emphasis on the negotiated approach to integrated river basin management; watershed development; ancient and traditional techniques of water management; and environmental law and practice, especially public interest litigation. His publications include *Evaluating the Tehri Dam* (1988) and *High Dams on the Narmada* (1990) and *Integrated River Basin Management: A Negotiated Approach* (2005). He has co-edited *Rehabilitation Policy and Law in India: A Right to Livelihood*. He is currently a member of the Expert Committee on Interlinking of Rivers, Ministry of Water Resources, Government of India and of the High-level Monitoring Committee appointed by the Ministry of Environment and Forests, Government of India, on the Mahabaleshwar-Panchgani Eco-sensitive Zone. He is also the Chairperson of Gomukh Environmental Trust for Sustainable Development, Pune.

Rama Prasad served as a Professor of Civil Engineering at Indian Institute of Science, Bangalore. He has a Bachelor's degree in Engineering from the University of Mysore and a Ph.D. degree from the Indian Institute of Science, Bangalore. He has published extensively in both national and international journals in the area of Hydraulics, Irrigation and Water Resources. Professor Prasad is a member of several professional bodies and has served as member in many technical committees set up by the government. He has served as an expert witness before the Cauvery Water Disputes Tribunal.

S. Settar is S. Radhakrishnan Visiting Professor at the National Institute of Advanced Studies, Bangalore and Honorary Director, Southern Regional Centre of the Indira Gandhi National Centre for the Arts, Bangalore. Professor Settar was awarded his Ph.D. from the University of Cambridge. He was formerly Professor of History and Archaeology and Director of Indian Institute of Art History (1971–96), Chairman of the Indian Council of Historical Research, New Delhi (1996–99). He has authored and edited several books in the fields of Art-history, Archaeology, History, Religion and Philosophy and Indian Classics. Three of these are published in Europe (Denmark, Holland and Germany). He is also the recipient of a dozen awards, including two national awards, one by the Indian History Congress and another by the Central Sahitya Akademy.

Index

adjudication, processes, 68–70; of water disputes, 85–86
Administrative Reforms Commission of 1969, 15
agriculture, 3, 7, 9, 39, 50–51, 53, 79, 130; Cauvery river for, 147; and development, 168; water-intensive, 58
Amaravati Project, 106
arbitration and negotiations, 14
Area Water Partnership, 137
arid areas, 23
artificial recharge. *See* rainwater harvesting
awards of River Disputes Tribunals, 91–92

Babhali Dam, 123
Baglihar Project, 77–78
Baikal Lake, 53
Basin-Level Platform, 135–36
Berlin Rules on Water Resources, 12
Bhima Partnership, 137
Bhima river basin, conflict resolution on, 129–30
Bhima Riverine Ecosystem, 133
Bhima system, using water of, 137
Brahmaputra, 161
Brahmaputra–Barak basin, 83

Carius, Alexander, 54
Castro (2005), on water, 172
Cauvery Fact Finding Committee (CFFC), 109–10
'Cauvery Family', 15
Cauvery Water tribunal, 143; Neutral Expert, decisions of, 77–78; order for allocations, 77, 115; report, 78

Cauvery/Kaveri River, 36, 99, 140–41, 163; of ancient Kannadigas, 101–3; of ancient Tamils, 99–101; award and violence, 123; basin, 108, 167; Bed Dam, 103; branches of, 118n; for city's drinking water, 52. *See also* urbanization; conflict and regional identities, 170; dispute, 13, 48; during colonial period, 103–5; during post-colonial period, 105; Fact Finding Committee, 107; and inter-state water disputes, 149; Mettur Project, 104; *Purana*, 100–101; rises in Western Ghats, 108; River Authority, 92; Water Dispute Tribunal, 107; waters, sharing of, 47
Central Ground Water Authority (CGWA), 89, 90
Central Ground Water Board, 87
Central Water Commission (CWC), 9
Centre for Science and Environment (CSE), 32
change: as conflict, 56–58; as opportunity, 58–61
check dams, 87, 132
civil society, for integration, 125–27; organizations, 16
community-based: institutions 28; knowledge, 178
conflicts, 48; and institutions, 54–56; problem of, 117; resolution, 36–37, 128–29, 166; root cause of, 79–80
conservation, 86–87
constitutional mechanisms, 37

Dabelko, Geoffrey D., 54
desert areas, 23
development: activists, 177; of hydropower, 86; sustainability of, 83–84
'Dialogue on Water for Food and Water for Nature', 137
Draft Maharashtra Water Resources Planning and Regulation Authority, 2002, 136
drinking water, 25–27; access to, 25–27; through tankers, 27
Drought Prone Areas Programmes (DPAPs), 31, 132
droughts, 83, 88
'dryland blindness thesis', 31

Economic Instruments, 95
ecosystem changes, 128; on marginalized communities, 127
effluence treatment, 24
Eighth Five-Year Plan (1992–97), 26
Environment Protection Act, 1986, 89
equitable: distribution, 130–32; utilization, 11

facilitator, role of, 153–54
Farakka barrage, 6
Flood Plain Zoning Regulation, 88
floods, 88
'Food for Work', 40
Forest Conservation Act, 1980, 89, 133
forest department and negotiations, 132

Ganga–Brahmaputra–Meghna system, 82
Ganges (India–Bangladesh), 71
Ganges, 6–8, 161, 163; and Farakka barrage, 6–7

Giordano, Mark, 55
Global Water Partnership (GWP), 157
Godavari river, 13, 163; Award, 123
Gokak agitation, 48
Gomukh, initiative of, 16
Gomukh Trust, 126, 130, 131, 134–35
governance of water, 14–15
Gram Sabhas, 130
Greater Mekong Subregion, 60–61, 63
Green Revolution, 51–52
ground water, 27–28, 114; aquifer, recharging, 87. *See also* rainwater harvesting; development in Nagappattinam, 147; extraction, 28; irrigation, expansion of, 27, *see also* irrigation; Legislation, 87–88; expansion of, 24; management, 135

Helsinki Rules, 10–12, 72
high-yielding varieties, 27
Himalayan rivers, 4, 161–62, 163

ILC, 11
India, 3–4; inter-state transboundary water sharing, 7; sources of water in, 4
Indian River Basins, 161–68
Indus (India–Pakistan), negotiations over, 71
Indus river, 161; to Pakistan, 6; systems, 7–8
Indus Water Treaty, 6, 77–78
Institutional Reform, 37
institutions, 61–63
Integrated Water Resources Development and Management (IWRDM), 83, 86 157–58, 177
Integrated Water Resources Management (IWRM), 18, 33–34

Inter State Council (ISC), 16–17
interactive approach, 76
Inter-basin organizations, 160
International River Basins, 158
inter-sectoral conflicts, 27
Inter-State Council (ISC), 15, 37
Inter-state River Water Disputes Act (ISRWDA), 1956, 85
Inter-state river water disputes, 47, 141; resolving, 72; tribunals, 75, 122; for water sharing, review of, 126
Inter-state River-Water Disputes (ISRWD) (amendment) Act, 85
inter-state transboundary water sharing, 7
Inter-State Water Disputes (ISWD) Tribunals, 70, 72, 79, 85
Inter-State Water Disputes Act (ISWD), x, 13–14, 36, 66–67, 73, 74, 76, 78–79, 154
irrigation, 7–8, 24–25, 106; and Egypt and Sudan, 160

Jeevan, 131

Kabini basin, 117
Kannada movement. *See* Gokak agitation
Kannambadi Dam, 104
Karnataka Watershed Development Project (KAWD), 31
Karnataka, for criticism, 77; release at border of, 110; and Tamil Nadu over Cauvery, 80; and Thanjavur farmers, 147
Kaveri. *See* Cauvery
Kaveri–Poompattinam. *See* Cauvery/Kaveri River
Kerala, 78, 99, 106–7, 108, 110, 117, 123, 140, 149
knowledge systems, Agarwal on, 176
Kramer, Annika, 54
Krishna river, 13, 165

Krishna Valley Authority, 92
Krishna–Godavari conflict, resolution of, 163
Krishnarajasagara Dam, 112, 147; construction of, 105. *See also* Cauvery/Kaveri
Kuruvai, 114, 118n, 147

legal arbitration, 121–22
legal reforms, 73–75
local knowledge systems, 176, 179
Lower Bhavani Project, 106
Lower Coleroon Anicut (LCA), 109–10, 112
Lower Mekong River Basin, 5

Madras Institute of Development Studies, 15
Madras–Mysore Agreement, 103–4; expiry of, 105
Maharashtra Farmers Participatory Irrigation Management Act, 2002, 136
Maharashtra Krishna Valley Development Corporation (MKVDC), 134
market-based mechanisms, 34. *See also* water markets
Measurement Error Freezing Problem, 116
Mekong: Basin, 59, 161–62; Committee, 59; River, 59
Mekong River Commission, 59
Mettur Canal Work, 106
Mettur Dam, 105; flow recorded at, 117. *See also* Cauvery/Kaveri
millennium development goals (MDGs), 172
Ministry of Environment and Forest (MOEF), 89
Ministry of Water Resources, 92
multi-stakeholder platforms at local level, 130
multi-stakeholders' dialogue (MSD), 144–45

Murray-Darling River Basin in Australia, 6
Mutha river 134; ecosystem of, 135
Mutha River Improvement Plan (MRIP), 134

Nariman, Fali, 76
Narmada Bachao Andolan (NBA), 36
Narmada river, 13, 163, water dispute, 76
National Commission on Integrated Water Resources Development Plan (NCIWRDP 1999), 4, 8, 23, 72, 84
National Flood Commission (Rashtriya Barh Ayog), 88
National Inter-State Water-Sharing Policy Statement, 68
National Water Board (NWB), 9
National Water Development Agency, 9
National Water Policy, 12, 67–68, 81, 84–85, 92
National Water Resources Council (NWRC), 9, 37, 68, 91
natural resources, disputes in sharing of, 145–47
negotiated approach to river basin management (NAIRBM), 127
New Kalatalai High Level Canal, 106
New Legal Instruments, 95
NGOs for drinking water provision, 39
Nile initiative, 160
Nile: river 5, 160; basin, 158–59; action plan, 160

Panchayat Raj institutions, 130
panchayats, 26–27
participatory: approaches, 92; decision-making, 135; process for resolving conflicts, 137–38
peasants, promotion of, 106. *See also* agriculture
percolation 'losses' in paddy irrigations, 113–14
Pondicherry, 106, 108
Poornaiah, Dewan, 103
population, 23
pricing of water, 4, 34–35
public interest litigation (PIL), 29
Puducherry. *See* Pondicherry
Pullamalai Canal, 106
Punjab Termination of Agreement Act, 2004, 123

rain-shadow region, 23
rainwater harvesting, 87. *See also* re-use of waste water
Rajamannar Committee of, 1971, 15, 17
Ravi–Beas case, 13
Rawls, John, 62–63
redistribution of resources, 172
rehabilitation, funding for, 29
reservoirs, 93, 113, 116, 130–31, 160
re-use of waste water, 87
riparian states of Karnataka and Tamil Nadu, 15
River Basin Organizations (RBOs), 14, 36, 37, 161, 91–92
river-basin: development and management, 124–25; partnerships, 136–37
River Boards Act 1956 (RBA), ix, 14, 66–67
River Interlinking Project (RIP), 28–30
river water: conflicts, 50; sharing and conflict resolution, 10

Sadat, Anwar, 52
Sardar Sarovar Project (SSP), 36
Sarkaria Commission of 1983, 15, 17

scarcity and conflicts, 50–54
Scheduled Tribes and water, 40
Self-help group, 130
Social Change Initiatives, 95
social dialogue, 143–45, 146–48, 149–50; objectives of Cauvery multi-stakeholder, 150–51
South Asia Water Forum at Kathmandu, 137
south-west monsoons, 23
Special Leave Petitions (SLPs), 78
storage projects, 42. See also tank
strategic action plan (SAP), 17
Supreme Court judgment, 141–43, 155n; violence over, 142
surface water potential, creation of, 28
surface-based systems, 24–25; irrigation projects, 24. See also irrigation
Sutlej–Yamuna Link Canal, 123

Tamil Nadu, 141; and Kerala, 107
tanks, 52; management institutions, 33
technology and crop yields, 25
technology-based water resource management, 177
TMC, 118n
traditional knowledge systems, 175–76; and alternative paradigm, 177–79
Traditional Water Management Systems, 32–33
transboundary rivers, cooperation over, 4–7; water sharing, 5, 10–18; water disputes, Aaron Wolf on, 54–56
Treaties of United States and Mexico, 8
Tribunals, 13–14, 77; and Helsinki Rules, 68

UN Convention of 1997, 10–11
UN Water Conference, 156

urban water demands, 24, 26. See also water use
urbanization, 23, 33, 52, 81, 168

Water (Prevention and Control of Pollution) Act, 1974, 89
Water (Prevention and Control of Pollution) Cess Act, 1977, 89
Water and Land Management Institutes, 9
water charges. See pricing of water
Water Dispute Act, 1956, 106
water disputes tribunal, 85
Water Quality Review Committees, 90
Water Resource Management, 169 176; actions need for, 93–94; Administrative Initiatives, 94; Policy Focus, 94; socio-political dimension of, 169–72
water resources development, environmental aspects of, 88–90
water sharing, patterns of, 168
'water stress', 5
Water Users Associations (WUAs), 25, 92
'water wars', 5, 36
water: availability, 82–83; and Constitutional provisions, 84; and efficient crops, 25; equity, 38; Institutions, 7–9; intensive technology, 58 (see also yield determination problem); markets, 34–35; and population growth, 83; quality of, 8; related disputes, 158–61; scarcity, 53, 173; sectoral usage of, 23–24; sharing arrangements, 53; as States' responsibility, 9
Water Quality Assessment Authority (WQAA), 89–90
Watershed Development Programmes (WDPs), 30–32

water-use, 7–9; urban and agricultural, 134–35
WEAP (Water Evaluation and Planning System), 151
Western science, 178–79
women: and water, 38–39, 173–75; and participation in the management of land and water resources, 174–75

World Bank, 26
world rivers, 4

yield determination problem, 115–16
Yoffe, Shira, 55

Zeravshan River Basin, 58

For Product Safety Concerns and Information please contact our EU representative GPSR@taylorandfrancis.com
Taylor & Francis Verlag GmbH, Kaufingerstraße 24, 80331 München, Germany

www.ingramcontent.com/pod-product-compliance
Lightning Source LLC
Chambersburg PA
CBHW070610300426
44113CB00010B/1477